INTERNATIONAL
BLOGGING

Steve Jones
General Editor

Vol. 50

PETER LANG
New York • Washington, D.C./Baltimore • Bern
Frankfurt am Main • Berlin • Brussels • Vienna • Oxford

INTERNATIONAL
BLOGGING

Identity, Politics, and Networked Publics

Edited by Adrienne Russell, Nabil Echchaibi

PETER LANG
New York • Washington, D.C./Baltimore • Bern
Frankfurt am Main • Berlin • Brussels • Vienna • Oxford

Library of Congress Cataloging-in-Publication Data

International blogging: identity, politics, and networked publics /
edited by Adrienne Russell, Nabil Echchaibi.
p. cm. — (Digital formations; v. 50)
Includes bibliographical references and index.
1. Blogs—Political aspects. 2. Blogs—Social aspects. 3. Blogs—Case studies.
4. Internet—Political aspects. 5. Internet—Social aspects. 6. Political participation.
I. Russell, Adrienne. II. Echchaibi, Nabil.
HM567.I56 302.23'1—dc22 2008024753
ISBN 978-1-4331-0234-9 (hardcover)
ISBN 978-1-4331-0233-2 (paperback)
ISSN 1526-3169

Bibliographic information published by **Die Deutsche Bibliothek**.
Die Deutsche Bibliothek lists this publication in the "Deutsche
Nationalbibliografie"; detailed bibliographic data is available
on the Internet at http://dnb.ddb.de/.

Cover design by Clear Point Designs

The paper in this book meets the guidelines for permanence and durability
of the Committee on Production Guidelines for Book Longevity
of the Council of Library Resources.

© 2009 Peter Lang Publishing, Inc., New York
29 Broadway, 18th floor, New York, NY 10006
www.peterlang.com

Printed in the United States of America

To
Michela, Adam, Mattia
John, Sofia, Sam

Table of Contents

Acknowledgments

The idea for this book came in 2005 when Nabil was teaching at Franklin College in Switzerland and Adrienne at the American University of Paris. At these international campuses, which draw undergraduates and faculty from around the world, we were experiencing first hand the ways communication technologies were being developed and adopted variously by people depending on their national and cultural contexts. Our experience with our colleagues and students mirrored a significant reality that we had been noting in the work of international new-media users and communities and online activists but that had received scant attention among media scholars.

The diverse expertise and engagement of the contributing authors made the book possible. Likewise, the guidance of Steve Jones, the series editor, and the work of Mary Savigar, Bernadette Shade, and Sophie Appel at Peter Lang energized the project and elevated the content. Adrienne thanks the University of Southern California's Annenberg Center for providing her the time to work on the book. She also thanks her colleagues there who inspired and advanced much of her thinking on the issues the book explores. The editors extend special gratitude to the following colleagues and friends who generously shared ideas on specific related topics or who commented on drafts of chapters or both, in particular Lynn Clark, Waddick Doyle, Corinna di Gennaro, Seeta Peña Gangadharan, Mohamed Hamidi, Mimi Ito, Merlyna Lim, Howard Rheingold, and Karim Zeribi. Finally we thank Michela Ardizzoni and John Tomasic, our most indulgent supporters and most candid critics. It is to them and to our sweet children that we dedicate the book.

CHAPTER ONE

Introduction

International Blogging—Identity, Politics, and Networked Publics

ADRIENNE RUSSELL

I

This book is part of an increasing effort in media studies to address the parochialism of contemporary scholarship by considering media practices and products developed throughout the world. In 2000, citing corporate-led globalization, developments in geopolitics, the rise of the Asian economy, the emergence of new centers of media production, and the growth of media studies as an academic field, James Curran and Myung-Jin Park (2000:3) called the narrowness of media scholarship "transparently absurd." In the years since, proliferation of new forms of digital media and the related rise of the audience as a major participant in the production of online content extends even further the range of media products and practices developing worldwide and the absurdity of theory elaboration based on isolated Western case studies.

The aim of *International Blogging* is to broaden consideration of blogging, a now major worldwide new-media practice, by emphasizing the significance of context. We have included contributions by scholars from nine countries that testify to the complex set of factors that shape national and language-based blogospheres. The book's case studies focus on blogs that originate outside the United States since a large body of work on American blogging already exists. Over the past decade, blogs have become a significant part of the transnational media environment, the most popular of so-called 2.0 or second-wave Web applications. Yet analysis of the form generally reflects the traditional limits of the field. The fact is new-media scholars by and large perceive and assess blogs around the world according to a particular perception of the form's qualities. This is an updated version of the mistake we made with old media. Early champions of the potential of television as a tool in international development, for example, imagined that TV everywhere would act among viewers the way TV does generally in the United States. But more recently communications scholars such as Elihu Katz and Tamar Liebes (1993)[1] have demonstrated, for example, that people watch TV and relate to TV in different ways, that they look for different things from it depending on cultural contexts and that they likewise make different meanings of the things they see. Blogs such as the popular one produced by "riverbend"[2] in Baghdad suggest blogging as a form may or may not be fostering political representative democracy in Iraq, but it is definitely strengthening traditional forms of communication there, such as oral-style micro storytelling, as the key ingredient to larger cultural conceptions.[3]

To many, the spread of the American blogging model around the world—including its norms and practices and modes of operation—effectively represents the spread of democracy. The rhetoric that surrounds blogging essentially describes the liberating potential of a new (American) cultural product, created and distributed globally through inherently democratizing digital tools and networks. More specifically, a rash of recent works outlines the emergence of a new more horizontal politics and journalism driven by blogs and the networks that blogs seem to engender.[4] These works mostly derive from compelling anecdotal evidence but also mostly overlook or ignore the ways power dynamics offline influence developments online. There remains generally a crucial lack of integration in new-media studies between online and offline realities. The theoretical links scholars have been forging, myself included, between democracy and the Internet generally and blogs in particular form the great bulk of popular as well as official thinking, obscuring variable contexts and hemming in larger realities.

The United Nations World Summit on the Information Society in Tunisia in November 2005 demonstrated the way this assumed relationship between technology and communication extends beyond the scholarly community.[5] The event was charged with the tension between the summit delegates, who were promoting increased access to and openness toward new media, and the Tunisian government, which was overtly repressing media freedoms across the board. There is a metaphorical quality to the fact that though the delegates in the summit hall considered the potential of the Internet to foster development and democracy, the people just outside the hall were living with the ability of the government to foster repression. During the summit, some of what was going on in Tunisia included a hunger strike staged by eight Tunisians over human rights violations, including indefinite detentions for posting or viewing "subversive" material online; visa authorities prevented the head of Reporters without Borders from entering the country; a human rights reporter from the French daily *Libération* was beaten on the street while nearby police refused to intervene; and state communications agents blocked or took down Tunisian protest and political Web sites. Defending the selection of Tunisia as the site for the conference, UN Secretary-General Kofi Annan told the BBC that he personally discussed the issues of censorship and human rights abuses with the country's president. "Sometimes," he said, "organizing these conferences in places like Tunisia, putting the spotlight on them, where these issues of human rights and others are discussed, it's extremely helpful, it helps push the cause forward" (BBC, 2005).

Annan's remarks reference for me the problem with the way a lot of scholarship and related bureaucratic writing imagines the power of digital media—the idea, that is, that digital media's just being somewhere "helps push the cause forward." In his remarks, Annan was talking on one level about drawing media attention to rights abuses and even about provoking oppressive heads of state. On another level, however, he was drawing on a theoretical tradition that promotes the benefits of exposure—of exposing people from oppressive countries to people and institutions from the lands where democracy reigns, the idea being that just getting the UN and democratic thinkers on the ground would create a force, born from example, that would move life in "places like Tunisia" closer to something like life in the West. This theory of exposure has thread its way through media studies for a long time, shaping academic research as well as government policy, and, I think, is enjoying a rebirth in communication and digital-media studies today, tied in particular to the world of blogging.

Recent writing on the liberatory potential of digital media constitutes the latest chapter in the promotion in the West of media as perhaps the key tool in the spread of democracy. Theories of international communication were an integral part of Cold War discourse—the primary function of international communication being, according to Western thinkers at the time, to promote democratic government, freedom of expression, and financial markets (Thussu, 2000). Cold War-era or second-wave modernization theory[6] arose from the notion that, in the global ideological battle against socialism, international mass communication could be used to transfer the social, economic, and political models of the West to the newly independent countries of the South. One of the earliest exponents of this theory, Daniel Lerner (1958), proposed that contact with the media facilitated societal evolution from "traditional" to "modern" because the flickering presentation of modern ways spurred members of traditional societies to reassess their ways of life.

Subsequent research thankfully "problematized" modernization theory, in part by breaking down the simple dichotomies at the heart of much of the writing. Case studies demonstrated that, despite tireless efforts at modernization through media, traditional cultures and values endured, even as people throughout the non-Western world adopted and adapted the latest communication technologies.[7] In the late 1960s, proponents of media imperialism theory directly challenged modernization-ists with the argument that American media aid to developing countries, rather than freeing people of the traditions alleged to inhibited development, created increas-ing dependency within the already imbalanced global economic system, serving mainly to spread Western consumer values and tastes around the world (Boyd-Barrett, 1977; Mattelart, 1979; Schiller, 1976).

International communication theory and research have developed a great deal of nuance in the decades since, underlining, for example, the way U.S. media prac-tices have been remade by users around the world to better suit their needs (Appadurai, 1996); the fact that people interpret media texts differently depending on, among other things, their gender and cultural identity (Ang, 1985); the gener-al preference for regional and national media over global media products (Tunstall, 2007); the fact that there are significant flows of media from the global south to north (Thussu, 2000); and that media technologies are as often used as tools of oppression as of liberation (Downing, 2000; Lim, 2003; Sreberny-Mohammadi & Mohammadi, 1994).

And yet something fundamental to the Cold War discourse has been reanimat-ed. Digital communication in general has been touted for its independence relative to mass communication, its lack of gatekeepers, its mostly unmediated network qual-ities (Negroponte, 1996; Rheingold, 2000; Turkle, 1997). Discussion of blogging takes this thinking to new levels. Blogging is celebrated as extended public journal-ing, pure multimedia freedom of expression, produced anywhere in the world there is Internet access and available for eyeballs the world over to take in. The democ-ratic character of blogging is accepted as inherent, the very essence of both the act and the product, the starting point of any larger discussion.

Blogs are seen as part of, even perhaps fueling, a trend toward more outspoken, unruly, and mobilized publics, even if the manner in which these publics are being received is accepted as highly contextual (Benkler, 2006; Jenkins, 2006; Russell et al., 2008). There is, at base, still a tendency to presume the existence of positive affinities rooted in artifacts. As Luke (2006) puts it: "Everyone on the Net alleged-ly wants unconstrained and free connectivity to something, but sharing access to, and the use of, a set of telematic tools may not automatically create a free and equal fraternity of meta-nationals" (175). On the contrary, a "global village" does not spon-taneously occur. To the extent that it exists at all in the blogosphere, the global vil-lage is engineered through projects such as the Harvard Berkman Center's Global Voices Online, a "citizens' media" outlet that aggregates blogs, translates content, rallies around persecuted authors, organizes conferences, and solicits funding and support.

Through aggregator sites, activist tool-kits that include "how-to" guides and financial support, blogging advocates such as Global Voices promote particular practices and serve as de facto gatekeepers of the blogosphere. In the *Handbook for Bloggers and Cyber-Dissidents* published by Reporters without Borders, Mark Glaser contributed a chapter titled "What Really Makes a Blog Shine" in which he compiles a list of recommendations for bloggers from countries where "the government is watching their words very carefully and the world is watching them as well." He suggests bloggers from these countries use their own voice, post frequently and about current events, involve readers in the conversation, and include "good old-fashion reporting." His suggestions are clichés indiscernible from those you will find in contemporary U.S. business or journalism textbooks on new media; they are in fact recommendations about how to make your blog palatable to American readers, extolling American-style individualism and drawing on traditional American news practices that insist on timeliness and truth-claims based on the observations of reporters and their sources.

Global Voices anoints "bloggers who shine" as what it calls "bridge bloggers," adding them to the ranks of its list of "the most influential or respected and credible bloggers or podcasters" from around the world. Global Voices aggregates these "A-list" blogs and serves as a sort of watchdog, reporting cases of persecuted authors, garnering international support for them, guarding them against isolation. Indeed, Global Voices is just about equally split in its content between bloggers who are bringing mainstream media to task or unearthing new journalistic information and those who have been the target of censors and/or jailers and whose fate is dependent on their ability to capture the attention of an international audience, beginning of course with popular bloggers and their readers.[8] The tension at the heart of laudable projects like Global Voices is the one at the heart of the blogosphere today, one so obvious as to go rarely remarked upon, a recognizably Western tension that runs between free speech and democracy on one side and marketing and public relations on the other. Clay Shirky (2005) in his popular essay "Power Laws, Weblogs, and Inequality", argues what most of us take for granted, that these inequalities are not a failure of the system but rather an inevitable side effect of freedom of choice: "In any system where many people are free to choose between many options, a small subset of the whole will get a disproportionate amount of traffic (or attention, or income), even if no members of the system actively work toward such an outcome." The organization of the blogosphere is largely shaped by the fact that some members have been actively working toward that outcome for some time. The blogosphere star system is evidence that digital networks reflect offline power dynamics, the same dynamics that gave rise to theory of media imperialism. In the blogosphere, as on the Internet more generally, new forms of gatekeeping have arisen and new sets of skills are becoming established practice, the pre-requisites for entree into the realm of those with power on the Web.

Yet bloggers around the world producing material for local and national audiences seem to be developing in ways that are distinct from the U.S. model. For international communication scholars, these authors and their products have much to say about what lies beyond the hedgerow of A-list bloggers, calling into question assumptions that form the base of much of what we read on blogging and by extension on global amateur or DIY media.

II

Each of the case studies included in this book addresses network-era questions about the uniquely universal-particular and global-local qualities of our variously digital realities. Two distinct but overlapping themes have emerged from the authors' work—the use of blogs, first, to negotiate and articulate identity and, second, to resist political pressures. We have organized the chapters according to how closely they address the two broad categories, identity and power. The first half of the book consists of case studies that highlight the ways bloggers in France, Russia, and in a generalized Muslim cultural space negotiate, maintain, and exhibit identities online. The case studies that make up the second half of the book more directly address the ground being staked out by bloggers in traditional political communication spheres in China, Israel, Italy, Singapore, Australia, and on the so-called Arab Street.

In the opening chapter Nabil Echchaibi analyzes the intersection between mainstream news media and blogging in France. By focusing in particular on the *banlieu* blogging of French of North African decent, he underscores the work young authors are doing to provide a "more accurate" presentation of their daily life for French readers and to negotiate their identity as French citizens within France. Bloggers tell vivid stories of social discrimination, cultural ghettoization, and poor schooling, for example, that are seldom reported in the mainstream media. In "Theorizing the Muslim Blogosphere: Blogs, Rationality, Publicness, and Individuality," Eugenia Siapera examines the impact of blogging on the dynamic process of contemporary Muslim identity construction. Her chapter looks at how blogs provide an alternative understanding of the ways Muslims engage today's caricature topics such as the West, modernity, secularism, and Islam. Kim De Vries, in her chapter, considers how bloggers express a distinct ethnic and cultural identity for an audience perceived as simultaneously local and international and the ways communication practices between the bloggers and their readers also reflect this perception. Her analysis of five blogs written from China in English suggests that contrary to concern that they serve primarily as another conduit for Western cultural imperialism, the blogs have helped to define a robust national identity and undermine stereotypical images. Karina Alexanyan and Olessia Koltsova analyze Russian use of LiveJournal.com, a popular American blogging and social networking site with more than 400,000 users—one of the largest aggregates of Russian-language activity online. Their analysis of angry and fearful reactions on the part of users to LiveJournal's recent licensing deal with a Russian media company demonstrates how the ability of technology to foster the emergence of new transnational cultures is tempered by traditional geopolitical concerns and the enduring cultural histories and identities.

Several authors also underline the fact that the effectiveness of blogging as a political tool varies from locale to locale. Axel Bruns and Debra Adams use IssueCrawler, a Web-mapping tool, to identify and plot the issue networks among Australian bloggers and related sites on a number of key political issues. They argue that, contrary to received wisdom, political blogs in Australia act nothing like political blogs do in America: rather than mixing with mainstream media, they remain almost entirely apart from more conventional forms of political coverage. Similarly, in examining blogging in Israel, Carmel L. Vaisman concludes that unlike the American and Arab experience, where local bloggers have had direct impact on the political system and traditional media, mainstream journalists in Israel largely ignore political blogging, prompting bloggers to make direct contact with politicians. In a chapter exploring politically influential Arab blogs, Aziz Douai leans his analysis toward events in Morocco, suggesting that blogging has been effective there because it has been set in the context of a solid social movement that mobilized the "Arab street."

The final two chapters provide material on the populist power of the form. Giovanni Navarria explores the rebirth as a celebrity political blogger of blacklisted Italian comedian Beppe Grillo. Navarria demonstrates that Grillo has well harnessed the power of the Web to promote innovative modes of political participation, but he warns that Grillo, and other leaders who have emerged outside the traditional channels of institutional politics are as prone to populism or demagoguery as they are to high-end democratic debate. Yasmin Ibrahim uses the 2006 elections as a case study to explore the impact of blogging on political discourse in Singapore. She argues that political blogs have had symbolic as well as "performative" impact, suggesting that they have worked to re-mediate the political landscape by constructing new forms of civic participation, thus disrupting and displacing the dominant discourses of the nation through personal narratives.

III

More than simply pointing toward the variability of blogging practice and product, the chapters in this book suggest that blogging, like digital communication more generally, is being conceptualized differently in distinct cultural contexts. A blog can be more things than we are presently imagining, a vehicle of democratic expression, yes, but also a means to revive tradition, to explore identity, to conduct public relations, and so on and so forth. By looking at local contexts, we can develop more nuanced assessments of how blogospheres variously serve communication needs, how they exist in relation to one another, where they exist apart as well as where they overlap and how they interact with other forms of communication in the larger media landscape.

In 1996, Sandra Braman and Annabelle Sreberny-Mohammadi (1996) considered how the Internet might influence international public discourse: "Today the internet genuinely—at least for the moment—offers autonomous production processes for those with the ability to surf it. The Net in fact may offer the opportunity for the creation of a public sphere or public spheres genuinely outside of the bounds of any single nation-state or organizational entity"(38).

More than a decade later, however, the globalized online public sphere is being shaped by many of the same factors that shaped the nation-based mass-media public sphere, the most notable factor being a persistingly tilted international cultural power dynamic. In the current environment, the global blogosphere is engineered by traditionally powerful groups of people in the developed world. The criteria set by these people from which to select sites for translation and promotion are being increasingly absorbed around the globe, ratcheting up the number of blogs that identifiably conform and pushing aside those that do not, paring down the blogosphere both online and in the mind.

But from the time of its nineteenth-century formulation, the notion of an overarching public sphere has always been problematic. Nancy Fraser's (1992) version, which describes a landscape of smaller "subaltern" public spheres that push back against dominant deliberations, seems a more viable way to envision a positive network-era reality. To arrive at a richer understanding of blogging in particular and of digital-era expression more generally, media studies scholars must work to move beyond the notion that communication practices and products should be valued according to the democratic values supposedly embedded within them. We should be attempting instead to develop theories of international communication that can see variation as something other than nonmodern and (therefore) nondemocratic.

Notes

1   Their work on cross-cultural readings of the American primetime soap "Dallas" is still an influential and persuasive example.

2   For more on "riverbend," see Aziz Douai's chapter on the Arab blogosphere.

3   On the oral versus writing cultures see, for example, Walter Ong's *Orality and Literacy: The Technologizing of the Word* (Methuen, New York, 1982).

4   *Blog: Understanding the Information Reformation (That's Changing Your World)* (2006) by Hugh Hewitt, *Crashing the Gates: Netroots, Grassroots, and the Rise of People-Powered Politics* (2006) by Jerome Armstrong and Markos Moulitsas Zuniga; *Blog!: How the Newest Media Revolution Is Changing Politics, Business, and Culture* (2005) edited by David Kline, Dan Burstein, Arne J. De Keijzer, and Paul Berger; *An Army of Davids: How Markets and Technology Empower Ordinary People to Beat Big Media, Big Government, and Other Goliaths* (2007) by Glenn Reynolds; and *The Blog Ahead: How Citizen-Generated Media Is Radically Tilting the Communications Balance* (2006) by R. Scott Hall.

5   According to the United Nations, 19,401 people participated in the event, representing 174 national delegations, 92 international organizations (like UNESCO or UNICEF), 606 nongovernmental organizations, 226 business entities, and 642 media outlets. The conference was one of the best-attended UN conferences of the decade. (http://www.one country.org/e173/e17306as_WSIS_Tunisia_Story.htm)

6   See, for example, Daniel Lerner and Wilbur L. Schramm (Fwd. by Lyndon B. Johnson), *Communication and Change in the Developing Countries* (East-West Center Press, Honolulu, 1967); Wilbur L. Schramm (ed.), *The Impact of Educational Television: Selected Studies from the Research Sponsored by the National Educational Television and Radio Center* (Chicago: University of Illinois Press, 1960); and Wilbur L. Schramm and Erwin Atwood, *Circulation of News in the Third World: A Study of Asia* (Chinese University Press, 1981).

7   See, for example, Annabelle Sreberny-Mohammadi and Ali Mohammadi's *Small Media, Big Revolution* (Minneapolis: University of Minnesota Press, 1994), for example, on how during the 1979 Islamic revolution in Iran, radical groups depended on audio-cassettes to promote their theocratic anti-Western ideology.

8   Global Voices is sponsored by Reuters, which often picks up stories from the site.

References

Ang, I. (1985). *Watching Dallas: Soap opera and the melodramatic imagination.* London: Methuen.

Appadurai, A. (1996). *Modernity at large: Cultural dimensions of globalization.* Minneapolis, MN: University of Minneapolis Press

BBC (2005, November 16). *Annan calls for digital bridges.* Retrieved September 2007, from http://news.bbc.co.uk/2/hi/technology/4443392.stm

Benkler, Y. (2006). *The wealth of networks.* New Haven, CT: Yale University Press.

Boyd-Barrett, O. (1977). Media imperialism: Toward an international framework for the analysis of media systems. In J. Curran, M. Gurevitch, & J. Woolacott (Eds.), *Mass communication and society,* pp. 116–135. London: Edward Arnold.

Braman, S., & Sreberny-Mohammadi, A. (Eds.). (1996). *Globalization communication and transnational civil society*. Cresskill, NJ: Hampton Press.

Curran, J. & Park, M. (2000). *DeWesternizing media studies*. New York: Routledge.

Downing, J. (2000). *Radical media*. New York: Sage.

Fraser, N. (1992). Rethinking the public sphere: A contribution to the critique of actually existing democracy. In C. Calhoun (Ed.), *Habermas and the public sphere*, pp. 109–142. Cambridge, MA: MIT Press.

Jenkins, H. (2006). *Convergence culture*. Cambridge, MA: MIT Press.

Lerner, D. (1958). *The passing of traditional society: Modernization in the Middle East*. New York: Free Press.

Liebes, T., & Katz, E. (1993). *The export of meaning: Cross-cultural readings of* Dallas. New York: Polity.

Lim, M. (2003). The internet, social networks, and reform in Indonesia. In N. Couldry & J. Curran (Eds.), *Contesting media power: Alternative media in a networked world*, pp. 273–288. London: Rowman & Littlefield Publishers, Inc.

Luke, T. (2006). Power and political culture. In A. L. Leah & S. Livingstone (Eds.), *Handbook of new media*, pp. 518–532. New York: Sage.

Mattelart, A. (1979). *Multinational corporations and the control of culture*. Atlantic Highlands, NJ: Humanities Press.

Negroponte, Nicholas, N. (1996). *Being digital*. New York: Vintage.

Rheingold, H. (2000). *The virtual community: Homesteading on the electronic frontier*. Cambridge, MA: MIT Press.

Russell, A., Ito, M., Richmond, T., & Tuters, M. (2008). Networked culture. In K. Varnelis (Ed.), *Networked publics*. Cambridge, MA: MIT Press.

Schiller, H. (1976). *Communication and cultural domination*. New York: International Arts and Sciences Press.

Shirky, C. (2003). Power laws, weblogs and inequality. *Writings about the internet*. Retrieved October 3, 2006, from http://www.shirky.com/writings/powerlaw_weblog.html

Sreberny-Mohammadi, A. & Mohammadi, A. (1994). *Small media, Big revolution*. Minneapolis, MN: University of Minnesota Press.

Thussu, D. (2000). *International communication: Continuity and change*. London: Oxford University Press.

Tunstall, J (2007). *The media were American*. London: Oxford University Press.

Turkle, S. (1997). *Life on screen*. New York: Simon & Schuster.

CHAPTER TWO

From the Margins
TO THE Center

New Media and the Case
of *Bondy Blog* in France

NABIL ECHCHAIBI

The Great Society created by steam and electricity may be a society, but it is no community
JOHN DEWEY (1927:98)

In the fall of 2005, as French mainstream media were struggling to gain access to the poor suburbs of Paris where young rioters were setting cars and buildings on fire, a team of heady reporters from a small Swiss weekly magazine were taking turns inside a cramped room of a suburban soccer club informally blogging about the riots. What began as a routine reporting mission by a Swiss reporter in the heart of the Bondy suburb in Paris unexpectedly turned into a prolific blog on life in the historically stigmatized French *banlieue* (suburb) and eventually became an epic experiment of participatory journalism in France. After two years, *Bondy Blog* has become a prime example of a networked form of communication whereby professional journalists and citizens from the banlieues collaborate to expose the realities and amplify the voice of a socioeconomic space that has been either caricatured or ignored by mainstream French media and politics.

Unlike other well-publicized international examples of bottom-up citizen journalism such as *Indymedia* or the Korean *Ohmynews* whose reporting is not limited to one specific location or group of people, *Bondy Blog* represents one segment of the French population it defines as disenfranchised and a victim of shortsighted political planning. From this unique niche position, the blog selects, discusses, and comments on political, economic, and social issues. Its ability to chronicle both the positive and negative sides of the banlieue and its role in the 2007 presidential elections, in which candidates made unusual campaign stops in the banlieues, have won the blog critical acclaim both in France and abroad.[1] Supporters of the blog argue that passing the baton of news production to citizens, or at least sharing it with them, has opened an unprecedented space for otherwise unsolicited voices to participate in public discussions and feel part of a heretofore exclusive French white identity.

In this chapter, I use the example of *Bondy Blog* in France to discuss the importance of acknowledging the role of political and social contexts in our analyses of blogging, bloggers, and blogs. As Bruns and Jacobs (2006) alert us, there are many uses of blogs, and talking about blogging as a uniform act with well-defined goals can be misleading in our attempts to determine its relevance and implications for both users and readers. Blogs remind us all of the waning gatekeeping powers of traditional media, but that does not necessarily translate into automatic empowerment or disenfranchisement for their users and readers. Much like all decentralized communication forms on the Internet, blogging has multiple functions, uses, and goals. Conflating these structural elements and styles without looking at individual cases is likely to undermine our ability to determine whether blogs can become liberating expressive vehicles for individuals and relevant political tools for communities.

Discussions of blogs have also been limited to established and recognizable blogs, also known as "A-list blogs," and this has ignored the important work of less-known bloggers who might better inform our conceptualizations of the role of blogging in an increasingly networked era. The focus of these discussions has been all too often polarized around two extremes: a celebration of blogs as a genuine public sphere tantamount to the eighteenth-century salons and reading cafés of prerevolutionary France or a dismissal of blogs as an amateur medium of biased information and reckless opinions. A-list blogs such as Andrew Sullivan's *The Daily Dish*, Joshua Marshall's *Talking Points Memo*, or Arianna Huffington's *The Huffington Post* may generate more news output and attract more readers, but they are different from smaller blogs because they rely more on the usual pundits already present in the mainstream media for news and comments. Less-known blogs not only incarnate better the potential of Internet users to challenge elite control of news production, but they also show us on a smaller scale how individuals and communities use network technologies to sustain new forms of social, cultural, and economic solidarity. My focus on *Bondy Blog* is an attempt to draw attention to "little-league" blogs, both in the United States and elsewhere, as equally viable sites to critically engage the shifting boundaries of communication and political power in the twenty-first century.

News Blogging: Shifting Power or Shifting Platform

Of all the Internet activity we have witnessed so far, blogging is perhaps the most paradigmatic of a digital revolution that has upset previous forms of traditional publishing and arguably tilted the power scale in favor of the individual at the expense of the old barons of content production. Some have compared the influence of blogs on widening the public discourse to the impact of the printing press on encouraging dissent during the Reformation (Bruns & Jacobs, 2006). Others have argued that blogs amount to a new form of journalism and as such may help this ailing profession find solutions to its dwindling audiences and retreating advertising revenues (Deuze, 2003; Deuze, Bruns, & Neuberger, 2007; Lasica, 2002). Though such deliberations on the redemption powers of blogs for journalism await more conclusive evidence, the impact blogging has had on old routines of news production and information control is undeniable. In the words of former Reuters editor-in-chief Geert Linnebank speaking at the 2007 Changing Media Summit, " . . . Everyone [today] can be a reporter, commentator or a film director . . . the days of owning and controlling these processes are over . . . the old value chain has been blown to pieces."[2]

As seen in emerging blogging practices around the world, old production structures continue to be challenged, rendering the consumption of information more and more complex as old content users become potential content producers. In an open source production mode, therefore, citizen participation is not only a democratic ideal but arguably a concrete tool that can be effectively leveraged to broaden the range of publicly discussed topics or even force the traditional media to heighten their news coverage of certain events or issues. Obviously, not all the output of existing participatory communication would sustain the argument that the Internet today is creating a responsible and engaged citizenry, but there are numerous attempts by some bloggers and other news Web portals to indicate the need has never been greater for alternative news sources that are free of institutional or corporate control over their reporting process.

Blogging may indeed have become the solution, or at least a panacea, to an increasingly privatized commercial culture, which James Carey (1995) argued has undermined an effective public sphere. Cyberspace could be the battlefield where traditional business models of the old media are supplanted by vibrant tones of public discussion and civic participation. A number of blogs today illustrate quite well the power of old audiences to compete with established media on breaking news stories, correcting false information, sustaining coverage of important stories, or simply starting an important dialogue on issues that get scant attention in public. In Egypt, where political freedoms are severely restricted, bloggers have been jailed for writing about the political grooming of President Mubarak's son to take over the leadership of the country or about the police indifference to mounting sexual harassment of young girls in the streets of Cairo.[3] The Egyptian blog, *Misr Digital*, is visited by thousands of readers for its chilling stories of police corruption and brutality rarely discussed in the mainstream media. In Myanmar, Ko Htike's blog was a rare window to see and read about the brutal government handling of the monks' pro-democracy demonstrations in the fall of 2007.[4] And in Iran, blogs such as *Rah-e Man* and *Daftar-e bi Mokhatab* have been one of the few places where Iranians could find public dissent of their president's nuclear standoff with the West.[5]

Consistent with philosophical conceptions of the ideal public as one that is inquisitive and willing to participate in discussions of public affairs (Dewey, 1927; Habermas, 1991; Williams, 1958), one can argue that blogging, particularly when it lives up to its full political potential as shown in the examples above, may be opening a public space in which individuals come together to deliberate on vital issues. But is this how blogging and other user-generated technologies are understood today by producers and users? Is there irrefutable evidence that the blogosphere is indeed a vibrant conversation likely to revive the public sphere? Or is the promise of new media technologies to further democratize political participation and civic discussion simply a utopian rhetoric? Will the Internet escape previous technologies' bias in favor of power centralization?

In his book, *Communication as Culture*, James Carey (1989) argued that communication must take us beyond the orthodox transmission model of information dissemination and into a ritual model in which the process of sharing information is revalued to benefit social and communal organization. Despite the ability of media technologies to privilege the type of grassroots dialogue Carey (1989) talks about, he cautions that such a desired result is not automatic and natural:

> Because we have looked at each new advance in communications technology as an opportunity for politics and economics, we have devoted it, almost exclusively, to matters of government and trade. We have rarely seen these advances as opportunities to expand people's powers to learn and exchange ideas and experience . . . The object, then, of recasting our studies of communication . . . is not only to more firmly grasp the essence of this "wonderful" process but to give us a way in which to rebuild a model of and for communication of some restorative value in reshaping our common culture. (34)

If we exhort Carey's concept of communication as a ritual in the era of blogging, we could say that potentially, and more than any other medium before it, the Internet and blogs might be creating a new network of media users who practice what Bruns (2005) calls "productive consumption." The audience today can strike back and produce its own networks of information that rival traditional sources of content production. Such a decentralized creation of information can be seen as revolutionary in its potential to change not only what we communicate but also how we communicate it, but intervening forces such as older media and consumer culture might tame this process and prevent it from reaching its full capacity in challenging any given social and political status quo. This does not mean that the political efforts of blogging might eventually be vain, but that the race for power control in this age of digital creativity will not be easy and certainly not solely determined by unique technological innovations. To temper the ongoing euphoric discussion of the importance of blogs, I believe it's best to cautiously approach this debate without exaggerating or belittling the impact of this phenomenon, as Williams and Delli Carpini (2004) argue in this statement:

> Optimistically, we believe that the erosion of elite gatekeeping and the emergence of multiple axes of information provide new opportunities for citizens to challenge elite control of political issues. Pessimistically, we are skeptical of the abilities of ordinary citizens to make use of these opportunities and suspicious of the degree to which even multiple axes of power are still shaped by more fundamental structures of economic and political power. In making our argument, we try to avoid the twin pitfalls of either seeing these changes as so profound and revolutionary that they fundamentally alter the political world or of seeing them as incremental extensions of age-old features of politics, hence signifying nothing new. (1209)

Indeed the popularity of blogs both in the developing and developed world is a clear indication that more people and communities embrace open source technologies to tell their own stories, rectify what they see as wrong, and inform others about their communities. The millions of blogs that populate the Internet and the thousands more that are created daily confirm how eager audiences of old media are willing to talk back. When blogging is added to the impressive diffusion of other interactive digital media like social networks, wikis, or podcasts, it becomes even harder to deny the fact that we live in a new era of complex and nonlinear communication. Much talk in the blogosphere, it is true, amounts to digital vanity: people who like to hear their own voices, but there is a small crowd of ordinary people whose blogging continuously shows that democratic publishing can be empowering to some degree.

With mixed intentions, such a realization has caused the mainstream media to revisit their closed production chain and find creative ways of engaging their audiences in all aspects of the reporting process including the selection and editing of news. After initially snubbing bloggers as an army of amateur journalist wannabes, well-established news organizations such as the BBC, Reuters, CNN, the *New York Times*, *Le Monde*, and many others now see an opportunity to reclaim their audience and remain relevant in the new media environment. As the blogosphere continues to grow globally at the rate of one blog per second, almost everyone in the traditional media today feature blogs created by their own reporters or open their Web sites for their visitors to suggest ideas for coverage or contribute their own information to stories. Some have reached even farther to accommodate the Internet's most dedicated users. In collaboration with Jay Rosen, one of the pioneers of citizen journalism, *Wired Magazine* started *Assignment Zero*, an ambitious open source reporting project to encourage Web users from around the world to tell a news story more completely while adhering to the highest standards of accuracy and truth.[6] In France, former journalists at the leftist newspaper, *Liberation*, founded in 2007 *Rue89*, a popular Web site that invites commentary on a range of political issues both in France and worldwide. Similar to *Huffington Post*, *Rue89*—the number is a reference to 1789, the year of the French Revolution—has outscooped French traditional media on various stories around the 2007 presidential elections.[7]

A more detailed description of the countless examples of conventional media adopting open source technologies is beyond the scope of this chapter, but it's worth noticing that the battle over information control has clearly shifted as ordinary people create their own content or are solicited for their information and comments. Participation in the news process whether through interactive old media or new media may never erode existing power relations between publisher and consumer, but as Manuel Castells (2007) says, the belief that you have power constitutes real power. Castells believes the new media is the new social space where power relations are challenged, and that interactive technology uses like blogging are making it easier for ordinary people to organize themselves as a decisive communication bloc. Both old media and mainstream politics are investing relentlessly in these communication technologies in an attempt to appeal to increasingly elusive audiences, and social actors can now use their blogs to influence the agenda of the old news media and wired politicians.[8] The outcome of this new framework, Castells says, is a reconfigured media space in which corporate media, political actors, and ordinary people share power.

There is, however, little evidence to conclude that the ascension of people's media to the realm of power will lead to concrete social change. The expansive reach of corporate media on the Web may indeed curtail the strident march of dedicated social actors to change the nature of public discussion and implement a real public sphere. But caution here does not mean we should relegate small bloggers and other disciplined Web activities to inconclusive experiments of angelic idealism. My work on *Bondy Blog* is an attempt to probe the political and journalistic relevance of a social group that had never had a representative public voice. It is also an attempt to assess the impact of old and new media convergence models on the future of news and its relevance for the public. Blogging, I argue, can create a viable public space for dialogue and political deliberation, but whether this space ensures more democracy might be an unfounded leap of argument at this point in time.

Blogging In France

In one of his first editorials as the new director of France's most famous daily, *Le Monde*, Eric Fottorino (2007) chastises French blogs and citizen journalism as "ad nauseam conspiracy theory" orchestrated by people who have no regard for accuracy and fact-checking. Few weeks after this highly debated editorial ran, *Le Monde* started its own official blog, *lepost.fr*, a classic example of citizen journalism where readers can submit their own articles and photos or work together with a site reporter to cover news events. *Le Monde's* contradictory move could be read as a desperate measure by an old newspaper to make a much needed transition to the new world of interactive news content. In fact, *lepost.fr* is a latecomer in a new media landscape in France, which has in the past few years seen a phenomenal exponential growth in the number of blogs and other alternative news Web sites.

Before the news Web revolution, one of the first blows to French contemporary journalism came from the popular free daily newspapers. Surveys in France in the last few years indicate a rising distrust with traditional journalists whose close relationships with French politicians, according to the public, have undermined their integrity and credibility. Younger audiences are also deserting mainstream news outlets in favor of the Internet and other free media. The press, about 4,000 publications serving national, regional and local audiences, is the hardest hit in France. Facing substantial budget cuts, newsroom layoffs, and plummeting advertising revenues,[9] old newspapers and magazines, both national and regional, fare poorly in readership compared with other industrialized nations. According to The World Association of Newspapers, only 167 daily newspapers are sold in France for every 1,000 inhabitants, compared to Japan (650), Great Britain (393), Germany (322), and the United States (263).[10] In 2004, *Le Monde's* readership fell by 4.1 percent; *Le Figaro*, a conservative daily, by 3.1 percent; and *Libération*, a liberal daily, by 7.8 percent.[11]

The rise of free dailies since 2002 has deepened the print media crisis in France, prompting warnings in the traditional press of "the dumbing down of journalism." Readership for such free dailies as *Metro* and *20 Minutes* have quickly surpassed that of well-established newspapers, taking away valuable advertising revenues in the process. Though news coverage of these "freesheets" might not be as in depth and sophisticated, the short writing style, the wider range of articles, and the emphasis on local issues are proving to be quite a successful formula for a new generation of time-poor and commuting readers.

Beyond this print competition, the news media in France are further beset by the rise of citizen journalism Web sites that follow the same model as the popular Korean *Ohmynews*. The editors of *Rue89*, a news portal founded in the middle of 2007 presidential elections by former journalists at the liberal daily, *Liberation*, believe their way of doing journalism reflects better how a new generation of content producers "wish to consume information: by being active, an actor who is heard and respected, and put on a par with professional journalists for a richer dialogue." *Rue89* has an average of 400,000 unique visitors per month, and that number is rising significantly, making it an important source of information likely to compete with traditional media.

Part of the quick success of sites like *Rue89* lies in the fact that the French spend a significant amount of time on the Internet daily. This is also why blogs have had notably more prominence in this country compared with others in Western Europe. In fact, France is one of the leading countries in the number of blogs in the world outside the United States. There are more than 6 million blogs in France, and close to 7 million French Internet users read blogs on a daily basis, according to Médiamétrie, France's leading media audience survey company. The number of blogs, according to the same survey, doubles every five months, and more than 4.5 million people have posted one or more comments on various blogs.[12] Loic Le Meur, France's pioneer and most prominent blogger, attributes this eagerness to participate in the digital public space to the fact that French people have always been outspoken about social, cultural, and political issues and that blogging has afforded them yet another platform for debate. His motto, "Traditional media circulate messages; blogs start conversations,"[13] has influenced a number of French people to start their own blogs. Loiclemeur.com is an extremely prolific blog in French and English where Le Meur, who is also executive vice president of Six Apart, a leading Weblog software company, blogs on politics and technology. The blog, which is read by 250,000 unique visitors per month according to Technorati, gained even more visibility and political importance when Le Meur officially endorsed Nicolas Sarkozy as a candidate for the 2007 presidential elections.

Consistent with the trend in the global blogosphere, French blogs are primarily personal diatribes of teenagers, but news and political blogs come second with a significant following. Political and news blogging in France became suddenly popular around the "No Vote" on the 2005 referendum on the European Union proposed constitution, which many critics described as former President Jacques Chirac's biggest political defeat. Much of the blogging activity during the government pro-constitution campaign was devoted to promote and mobilize support against ratifying the new constitution. The 55–45 "No Vote" was received as a surprise in domestic and international political circles, particularly after a massive "Yes" campaign by the French mainstream media. But the real debate about the implications of a uniform European constitution was in the blogosphere, where a Marseilles law teacher set up a blog to express his concerns with an EU document that grants further powers to institutions and leaves little discretion to individuals. His blog, which received 25,000 hits a day leading to the referendum vote, encouraged hundreds of blogs to do the same, prompting positive observations in France and abroad about the grassroots power of blogging and its relevance in the political process (Anderson, 2005). The relevance of blogging in France became even more apparent during the 2007 presidential elections. The candidates' decision to keep highly interactive blogs was seen as an attempt to avert a flop similar to the Yes-Vote campaign on the European constitution. Politicians also had valuable data to convince them they should move their campaigns to the virtual world. Data such as 27 million French are on the Internet daily and that 10 percent of this number keep well-read blogs pushed Sarkozy to team up with blogger Loic Le Meur in order to tap into this elusive audience. Le Meur even convinced Sarkozy to open an island on the popular virtual reality site, Second Life.[14] Christophe Grébert, a prominent local blogger on the left political spectrum in the city of Puteaux, was active during the elections exposing what he called the threat of Sarkozy's privatization of age-old public services in France. Grébert believes his blog, *monputeaux.com*, has gained him enough exposure to help him win his district seat at the 2008 municipal elections. Another milestone development in the short history of blogging in France was the 2005 suburban social unrest, during which some bloggers were praised for providing important inside coverage the traditional media were not capable of producing while others were vilified for inciting further violence.[15] In fact, the riots have revealed a new meaning of the role of blogging: for example, banlieue blogs were a new platform for marginalized suburbanites to comment on life on the socioeconomic fringes of French society. These blogs have laid bare a disturbing reality about the dangers in the French suburb and the lack of contact between people who live there and mainstream society. The blogger at *Parisbanlieue.blog* laments the fact that French journalists were paying 150 euros to talk to any presumed rioter, hoping to cover this story as a routine crime and delinquency event. He further blames French media and officials for continuing to depict the banlieue as an alien space:

> No, Paris is not burning, let tourists know so they keep coming. It's not a big deal, it's the banlieue, not Paris that is on fire. Minimizing what happened in the banlieues in the last few days reflects fear: that the riots would reach the capital, the beautiful Paris, and that bad images will circulate outside. . . . What a strange representation of the city and what segregation is behind this! On the one hand, it's the City of Lights, the sanctuary of the elites, of knowledge, of good taste and of intelligence. On the other hand, it's the banlieue, the renegade, the poor, the degenerated, the savage, and the "scum."[16] Shall we end up by closing down Paris to protect it, a toll way for cars, an ID card control for the people who live here? (2005)

Unlike *Parisbanlieue* and many other blogs of the suburbs, *Bondy Blog* was initially the sole initiative of an established magazine that saw in blogging a more appropriate way to tell a complex story about the banlieue that could not be told in a few newspaper columns or short television reports. Even in hindsight, the riots are still remembered as an ethnic and in some cases religious uprising, not as a statement about the social marginalization of the banlieue. This flagrant misperception is what prompted a Swiss magazine's experienced reporter to start a blog to give a more nuanced image of this urban space. A brief history of the blog will reveal the challenge the news site has had both within the banlieue from where it has been reporting and with French traditional media who accused the magazine bloggers/reporters of opportunism and lack of professionalism.

Bondy Blog: Brief History

Swiss magazine *L'Hebdo*, a small weekly with a circulation of 44,000, decided their chronicles of the unrest at the peak of the riots in Paris' suburbs were just too cliché and amounted to a marginal representation of the motives behind the violence. Serge Michel, a world affairs reporter at *L'Hebdo* at the time of the riots and now a war reporter with *Le Monde*, started a blog in the banlieue of Bondy, a 20-minute train ride from Paris, but a world away in urban infrastructure and economic opportunities. Michel believed the best way to understand the complex roots of the problems in the banlieue was through full immersion: "to live somewhere, to wake up in the morning and see these areas really in front of our eyes was important" (Boucq, 2006). Nine journalists from *L'Hebdo*'s different beats joined Michel in a tiny room rented from a local soccer club and worked on the blog during and weeks after the riots had ended in rotations of seven to ten days. They wrote about the people of Bondy, the homeless mother, the Saturday night parties, the unemployed youth, the young entrepreneur, the small grocery store, the dilapidated buildings, the local video rental, the travel agency, the day care center, the school district, the Internet café, and generally exposed realities their magazine couldn't afford to publish, given their small size.

The reporters/bloggers were generally welcome in Bondy and on many occasions were provided unique access to people and places other journalists working for conventional media may never reach. These encounters with the locals later resulted in a close working relationship, which later culminated in handing the blog to a team of young "bondinois." Indeed, in February 2006, the management of *L'Hebdo* decided, after inviting young bloggers to their newsroom for a crash course in journalism, to start yet a new experiment: give the blog to people who need it the most to continue to tell the ill-told story of the banlieue. *Bondy Blog* hired an editor, a high school economics teacher from Bondy, and a group of dynamic bloggers eager to become major players in their own space. Since the transition, the blog has filled an important lacuna in mainstream media coverage of the banlieue by serving as a constant open window. The posts may read less as journalism, but the topics they raise and the people they feature in their articles are unprecedented, as Nordine Nabili, the blog's current chief editor, says:

> I would challenge anybody to find 10 percent of the topics *Bondy Blog* covers anywhere in the French media. The banlieue is simply a story of crime, delinquency, school problems, and unemployment. We go beyond this façade to talk about the root causes of these issues. (N. Nabili, personal interview, author's translation, October 7, 2007)

Bondy Blog has become a popular source of information for curious ordinary newsreaders as well as for reporters whose access in the banlieue may be limited. The blog's coverage in this sensitive area cannot be ignored by reporters whose best access is still provided by hired fixers in the area. In fact, the blog is currently trying to replicate the Bondy experience in other city banlieues plagued with the same stereotypical images and marginal coverage in Lyon, Marseille, and Lille. Occasionally, bloggers from these banlieues contribute news coverage and commentary on issues affecting their own communities. Eventually, *Bondy Blog* hopes to create more local blogs elsewhere and produce a national blog representing a greater number of banlieues in France. It would be premature at this point to evaluate the success and impact of the existing local blogging initiatives, but the fact that other established media such as *Liberation* have recently opened *libevilles*, a network of local blogs targeting readers in the cities of Lyon, Toulouse, and Lille, might be a clear indication other media are taking their cues from the experience of *Bondy Blog*, as Serge Michel, the founder of the blog says here:

> At the beginning, the reaction [of mainstream media] was very enthusiastic because my colleagues did not see that the project of *Bondy Blog* meant the birth of a new medium. They thought it was simply a project to help the banlieues and the youth who live there, and after the riots of November 2005, everybody was ready to help the banlieues. Later, they realized they were in competition. The daily *Liberation*, for instance, is opening local blogs and also *Le Monde.fr*, with an emphasis on citizen journalism. And the relations between these two publications and *Bondy Blog* have become much more fresh. (S. Michel, personal interview, author's translation, October 15, 2007)

The budget of *Bondy Blog* is negligible compared to other newspaper blogs like the ones Michel refers to, but recently the blog has drawn interest first from *Yahoo France* and later from one of the most well-read free newspapers and online news sites, *20 Minutes*. The partnerships with these platforms may not render the blog a full-blown news operation at a large scale—20,000 euros from *Yahoo* and another 20,000 from *20 Minutes*—but the money helps cover the reporting fees of their bloggers who get about 20 euros per post. Based on this partnership, some of the posts published on the blog can appear in both the print and online publications of *20 Minutes*, giving coverage of the banlieues more visibility and much needed exposure. Other expenses are covered by the proceeds of a book on the blog Serge Michel wrote in 2006. Since its partnership with Yahoo, *Bondy Blog* has been receiving 6,000 unique daily visitors to their site, making it an important part of the French blogosphere, particularly considering the number of staff—a dozen of bloggers—and the size of their budget.

Bondy Blog is indeed viewed by its staff as an obligation to tell the story of people who have been denied a presence in French mainstream society and media. The bloggers are mostly young people of different ethnic backgrounds who reflect the ethnic and cultural diversity of the Bondy banlieue (with a population of 50,000), in particular, and the banlieues (with a population of 12 million) in France, in general, that have for years become an urban exile for immigrant families and working-class communities. Unlike the tiny number of teenagers who staged the riots of 2005 as a social and political statement about their marginal place in French society, *Bondy Blog* bloggers, who are high school students, disillusioned unemployed youth with degrees, or simply Bondy residents concerned about the economic degradation of their suburb, believe telling the story of their banlieue and its people through a blog might be more effective than burning cars and buses. Both the new and old staff of the blog believe the rioters in 2005 resorted to violence because they lacked a forum where they could express their anger and disillusionment with the political system that has neglected them, as Paul Ackermann, one of the early Swiss journalists who worked at the blog during its initial phase, says here:

> Before, I had the same view of French banlieues as anybody else from outside. I saw images of fire and it looked like war, too dangerous and violent. Once there, it was all different because we were learning to know people. The burning cars, buses, and buildings were the result of a malaise; not the beginning of a civil war. They were acts by youth revolted by the system, an epidemic reaction not legally acceptable, but these kids didn't know how to express their frustration otherwise. (Huynh, 2007)

The frustration Ackermann refers to is the aggregate of a long and precarious social experience in the French banlieue that has been heavily caricatured in the media. The teenagers of the banlieue, most of whom are from immigrant families of North and West Africa, have been consistently stigmatized as thieves, street gangs, brutal rapists (following rape incidents in the last few years), or lazy individuals who have no ambitions in life. The banlieues are also falsely viewed as a breathing ground for fundamentalist Islam since there is a heavy concentration of Muslim families in these urban settings. The 2005 rioters were in fact quickly labeled culturally alienated illiterates or Muslim delinquents whose sole objective was to set up a stronger Muslim presence in France (de Wenden, 2005). But the real problem of the banlieue, as it has become clear since the riots—and the work of

Bondy Blog may have contributed to this change in perception—is of a socioeconomic order and not a cultural or religious one. Many second, third, and fourth generation immigrants gradually rebuked their cultural heritage by refusing to be pigeon-holed in an ethnic group; some went as far as changing their names from Mohamed to Momo, Ali to Alain, and Rachid to Richard to find more competitive jobs or simply rent an apartment. This is the social context against which the bloggers at *Bondy Blog* have been writing their posts, giving rise to a new medium that is not a supplement of old media but rather a combination of old and new media.

Bondy Blog: The Audience Strikes Back

Since the riots, *Bondy Blog* has found in its autonomous platform an empowering tool to break conventional publishing barriers that have historically prevented the people of the banlieues from representing themselves to a larger French public. The nature of their reporting as an insider's chronicle of life in the banlieue provides a sense of credibility as well as a narrative of counter power, to borrow Castells's (2007) idea of social actors mobilizing themselves through alternative media "to challenge and eventually change the power relations institutionalized in society" (248). The ability to tell your story on your own terms in this case reflects a capacity to challenge a social order by countering its exclusive rights to define your image and identity. *Bondy Blog* posts might not garner the same attention as a daily newspaper article, but the fact that a largely ignored social reality in France is told for the first time on a consistent and accessible platform legitimizes the mission of the blog and validates its progressive role as a space for alternative voices to take on institutionalized power relations in French society.

One of the most enduring implications of the 2005 riots was the ascendance of the banlieue as an important political issue during the 2007 presidential elections. In an unprecedented way, political candidates were racing to get photo ops in the banlieue making promises of urban rebuilding and equal economic opportunities. During the campaign, *Bondy Blog* featured a high number of posts about the economic and social challenges of the banlieue and organized round tables and video sessions with various politicians, including the defeated left candidate Ségolène Royal, to determine and evaluate the political commitment to the banlieue. Making the campaign visible on the blog made the political process, long considered detached and negligent, relevant to the banlieue, and this, along with other civic engagement efforts in these areas, may have contributed to a record voter registration in the suburbs during the campaign.[17] The relevance of the banlieue during the elections has forced political candidates to propose concrete solutions to solve the many structural problems of the banlieues such as education, health, infrastructure, equal employment, and so on. This has also made traditional media more aware of their marginal treatment of the banlieue as a space to cover, as Michel confirms here:

> I think traditional media have improved their work on the banlieue, particularly since it became an important political issue during the presidential campaign and an issue on which to judge Sarkozy, who promised some sort of a "Marshall Plan" of the banlieues. Radio France, for instance, has opened one or two positions for correspondents in the banlieues. They called them RER, Reporters En Résidence. (Michel, personal interview, author's translation, October 15, 2007)

Following the success of their coverage of the presidential elections, the blog has teamed up with the free daily *20 Minutes, Canal/obs, Lemonde.fr,* and other media to cover the 2008 municipal elections in France from a banlieue perspective. The impact of this political coverage may not amount to a considerable overhaul of the economic situation of the banlieue, but it helps situate a heretofore-silenced community in the political map of France. In fact, the Marshall Plan Sarkozy promised during his campaign is still awaiting execution, and many posts in *Bondy Blog* that vividly depict the problems of the banlieues have been critiquing the time lag between political promises and delivery.

Beyond big events such as the elections, *Bondy Blog* provides a telling picture of day-to-day happenings in the banlieue without necessarily stereotyping the place as crumbling and dysfunctional. Some posts feature local artists, writers, and other professionals, while others provide portraits that humanize the ordeal of illegal immigrants in the suburbs as politicians debate a new immigration law that will make the use of DNA tests mandatory to verify the bloodlines of prospective immigrants who would like to join family members already living in France. The numerous comments these posts generate are also an indication the blog is creating a small community of readers who spare no verbiage to make their opinions known. Some posts can produce as many as 303 comments such as the one below about the failure of the French government to recognize the Paris massacre of October 17, 1961 in which perhaps more than 200 Muslim Algerians were killed and thrown in the Seine River:

> It seems in fact that a troubling page of history has been eliminated, torn and thrown into the fire to get rid of a shameful episode. . . . Nowhere in any school's curriculum does this episode appear. What's the position of the French government? What does it want? It's not until 40 years later in 2001 that the mayor of Paris Bertrand Delanoë takes the initiative to put a commemorative plaque honoring the numerous victims killed in this place. . . . France dares to give lessons to Turkey while it would be better if it reconciled its own history. (Kaddour, 2006)

The interactive component of blogs whereby readers can become instant commentators is what makes *Bondy Blog* relevant in a larger political spectrum. For the first time, Bondy residents and outsiders virtually converge on the blog to exchange points of view and simply instigate a dialogue that was rarely available before. This open channel of communication between these two communities might not readily cancel out misperceptions about the banlieue and its people, but it is a good starting point, a launching pad to secure a flow of necessary debate. It is still early to determine at this point whether this kind of public space constitutes a genuine public sphere that might eventually influence social relations and public policy. As Mohamed Hamidi, *Bondy Blog*'s former editor-in-chief and currently the director of *L'école du blog*, says:

> I wouldn't say our blog is changing the banlieue at this point. We're not having a direct effect on the mentalities of people here, but we certainly have changed the way the banlieue is heard. We simply didn't have a voice in public; now we do and it's a good strong one as well . . . we are also showing a different face of the banlieue, of young people who can take initiatives, who can debate political issues, and who can volunteer their time for a good cause. But I wouldn't say we're changing the banlieue. The banlieues need a lot of work. (Hamidi, personal interview, author's translation, October 10, 2007)

Such a deliberation does not necessarily mean that blog talk is necessarily empty air. The fact that *Bondy Blog* enables disenfranchised individuals to represent themselves and their neighborhoods is in itself a potent statement about the degree of these people's civic engagement. Some might argue that the capacity of creating our own messages in blogs only gives the semblance of civic involvement because much of that talk might be irrational or simply disorganized to constitute genuine constructive debate. The apparently emotional post above about France's colonial crimes might be irrational in that it lacks historical analysis and sophisticated proof of evidence, but does that mean that this kind of blogging debate is doomed because it simply doesn't live up to the ideals of rational debate? In criticizing Habermas's ideal public sphere, Francois Lyotard (1984) suggested disagreement, individuality, and even anarchy in discourse might guarantee a democratic emancipation. Any debate that requires consensus, he says, is likely to reify the limitations of meta narratives as the only viable framework to explain things.

As noted before, the posts of *Bondy Blog* can be imperfect and lacking in precision reporting, but the fact that there is a cadre of experienced journalists behind it indicates blogging is becoming less of a threat to some media professionals who have wised up to its potential as an adjunct to conventional journalism. Under this emerging new media paradigm, old media reconnect with deserting audiences, and social actors find new meanings in telling their stories using a new medium. Where is the convergence in this new framework? In the case of *Bondy Blog*, convergence takes different forms. First, the blog was launched by a journalist who continues to be its director and adviser. The current chief editor of the blog, Nabili, is an established radio journalist who worked for well-known radio stations in France. The blogging staff has been trained on various occasions at the newsroom of *L'Hebdo* and at the blog at the hands of working journalists. Recently, *Bondy Blog* has opened "L'école du Blog" (Blog School), a training program run by journalists and bloggers to help young individuals become more civically engaged in their blogging practices. And finally, *Bondy Blog* signed an agreement with the School of Journalism at the University of Paris to allow students to cover banlieue stories for the blog.

This convergence is an example of the coexistence of an autonomous blog with mainstream media. Far from being competition, partners in this case benefit from each another. In fact, these kinds of networks can help traditional media retain their relevance within a new media order in which user-generated content is as much horizontal as it is vertical. By opening its pages to *Bondy Blog* posts, outlets such as the daily *20 Minutes* afford its readers a unique local perspective on the banlieue produced by people whose only motive in covering this news is to make it known, not to be paid. We are living in new communication order, Castells (2007) says, in which old media have understood the imperative to "enter the battle in the horizontal communication networks" (259). In the case of *Bondy Blog*, therefore, the smaller and local posts of the blog representing a marginalized group will be irrelevant because in reaching social consensus these minority opinions and issues will simply be lost.

Old media are indeed beginning to slowly understand that in certain cases social network media thrive because they are—or appear to their users—genuine in their attempt to connect with people, that is they are free of the corporate attachments old media have been known for. Blogging can help boost the credibility of traditional media outlets, but they have to be careful how they tread in this space where users as much as old producers wield a significant power in controlling the flow of information.

It would also be naive to pose blogging and other alternative Web portals as a uniform solution to the current woes of conventional journalism. The bottom-up approach of blogging remains at variance with the gatekeeping model of traditional journalism, but the success of any convergence of these two media lies in their capacity to customize ways in which they can work together not replace one another. Blogging may also expand the spaces individuals and communities have for political deliberation, and journalists, besides incorporating these new spaces in their daily reporting as viable venues for their information gathering, should also be seeking underrepresented social groups and spaces on their own.

We need more case studies of small blogs from various international contexts in order to better evaluate the role blogging might play in the so-called digital democratic revolution. I have argued that virtual platforms like *Bondy Blog* can indeed serve its bloggers to positively rechannel energies and frustrations with a marginalizing political system, but for our research efforts to be more conclusive, we need to look at the implications of these decentralized information networks and analyze their potential to change political structures. With the exception of a few sporadic studies, so far evidence of the power of blogging on political change remains woefully anecdotal.

Notes

1. The *Bondy Blog* has won journalism awards in France and members of the staff have been invited in various countries to talk about their experience with the blog.
2. Guardian Changing Media: Reuters looks at the changes for "old media" accessible at http://strange.corante.com/archives/2007/03/22/guardian_changing_media_reuters_looks_a t_the_changes_for_old_media.php
3. BBC, http://news.bbc.co.uk/2/hi/middle_east/6164798.stm
4. http://ko-htike.blogspot.com/
5. http://news.bbc.co.uk/2/hi/middle_east/4650154.stm
6. *Assignment Zero Project* lasted about 12 weeks and folded because it proved that crowd reporting can be tough to manage. As one of the contributors later said, the project "suffered from haphazard planning, technological glitches and a general sense of confusion among participants." For more on Assignment Zero read: King Robert (2007) *Did Assignment Zero Fail? A Look Back, and Lessons Learned*. Available at: http://www.wired.com/techbiz/media/news/ 2007/07/assignment_zero_final
7. *Rue89* started in the second round of the presidential elections in 2007 with a scoop that the wife of Nicolas Sarkozy was not intending to vote in favor of her husband.
8. Countless politicians are creating Web sites and blogs since Howard Dean started his.
9. Pas de "Libération" dans les kiosques: http://www.rfi.fr/actufr/articles/071/article_39982.asp
10. France ranks 31st worldwide in daily newspaper readership. (EuroPQN)
11. French newspapers sales continue to plummet http://www.expatica.com/actual/article.asp? subchannel_id=25&story_id=17993
12. 15 blogueurs leaders d'opinion sur la toile: http://www.lemonde.fr/web/articleinteract- if/0,41-0@2-651865,49-759105@51-759106,0.html
13. Loiclemeur.com/france
14. French Politics in 3-D on Fantasy Web Site: http://www.washingtonpost.com/wp- dyn/content/article/2007/03/29/AR2007032902540.html
15. Banlieues: la guérilla dans les blogs tsr: http://www.tsr.ch/tsr/index.html?siteSect=200001& sid=6220743&cKey=1131382109000
16. Scum or "racaille" in French is a word then government minister, Nicolas Sarkozy, used publicly to refer to the rioters. It was later used against him to show his complete disconnect with life in the banlieues.
17. For more on how voter registration has increased in the banlieues during the 2007 presidential elections, read "Les jeunes de banlieue prêts à voter," http://www.afrik.com/article10932. html

References

Anderson, K. (2007). *Guardian changing media: Reuters looks at the changes for "old media."* Retrieved October 9, 2007, from http://strange.corante.com/archives/2007/03/22/guardian_changing_media_reuters_looks_at_the_changes_for_old_media.php

Anderson, K. (2005). Bloggers take on European elites. Retrieved September 10, 2007, from http://news.bbc.co.uk/2/hi/europe/4603883.stm

Boucq, I. (2006). *Blogging Takes Root After French Riots* Retrieved October 9, 2007, from http://www.voanews.com/english/archive/2006-04/2006-04-07-voa13.cfm

Bruns, A. (2005). *Online "produsers" dish up the news.* Retrieved September 28, 2007, from *ONLINE Opinion* Web site: http://www.onlineopinion.com.au/view.asp?article=3333

Bruns, A., & Jacobs, J. (2006). *Uses of blogs.* New York: Peter Lang.

Carey, J. (1995). "The Press, Public Opinion, and Public Discourse", in T. Glasser & C. Salmon (eds.) Public Opinion and the Communication of Public Consent. New York: Guilford, 373–402.

Carey, J. (1989). *Communication as culture: Essays on media and society.* Boston: Unwin Hyman.

Castells, M. (2007). "Communication, Power and Counterpower in the Network Society." *International Journal of Communication.* 1, 238–266.

Deuze, M. (2003). The web and its journalisms: Considering the consequences of different types of news media online. *New Media & Society, 5*(2), 203–230.

Deuze, M., Bruns, A., & Neuberger, C. (2007). Preparing for an age of participatory news. *Journalism Practice, 1*(3), 322–338.

Dewey, J. (1927). *The public and its problems.* New York: Henry Holt and Co.

Fotorino, E. (2007) Indépendance Retrieved August 9, 2007, from http://www.lemonde.fr/cgi-bin/ACHATS/acheter.cgi?offre=ARCHIVES&type_item=ART_ARCH_30J&objet_id=997573

Habermas, J. (1991). *The structural transformation of the public sphere: An inquiry into a category of a bourgeois society.* Cambridge, MA: MIT Press.

Hamidi, Mohamed. Personal INTERVIEW. 5 November 2007

Huynh, K. (2007). Paul ou la révélation des banlieues. *Bondy Blog,* October 30. Retrieved on October 30, 2007, from http://20minutes.bondyblog.fr/news/paul-ou-la-revelation-des-banlieues

Kaddour, H. (2006). Massacre du 17 octobre 1961 à Paris: Histoire d'un négationnisme d'Etat. *Bondy Blog,* October 18. Retrieved on October 29, 2007 from http://20minutes.bondyblog.fr/cgi-bin/display_index.pl

Lasica, J. D. (2002) "Weblogs: A New Source of News", in J. Rodzville (ed.)We've Got Blog, Cambridge: Perseus Publishing, pp. 171–182

Lyotard, F. (1984). *The postmodern condition.* Minneapolis, MN: University of Minnesota Press.

Michel, Serge. Personal INTERVIEW. 7 April 2007

Nabili, Noreddine. Personal INTERVIEW. 9 February 2007

Wenden, C. W. (2005) Reflections "À Chaud" on the French Suburban Crisis Retrieved September 7, 2006, from http://riotsfrance.ssrc.org/Wihtol_de_Wenden

Williams, R. (1958). *Culture and society, 1780–1950.* New York: Columbia University Press.

Williams, R. (1966). *Communications.* London: Chatto and Windus.

Williams, B. A., & Delli Carpini, M. (2004). Monica and Bill all the time and everywhere: The collapse of gate keeping and agenda setting in the new media environment. *American Behavioral Scientist, 47*(9), 1208–1230.

CHAPTER THREE

Theorizing
THE Muslim Blogosphere

Blogs, Rationality, Publicness, and Individuality

EUGENIA SIAPERA

The ongoing interest in the development and role of the blogosphere has led to several books on the subject (e.g., Burns and Jacobs, 2006; Tremayne, 2007). The important insights offered by these books, however, have not accounted for the parallel explosion of blogging by Muslims. Indeed, the popularity of Muslim blogs is well documented in both news media and academic publications, and this body of work has advanced fascinating arguments regarding the importance of, and functions served by, Muslim blogs. Typically, this work considers blogs as reflecting the dilemma of public versus private (Alexanian, 2006), as contributing to the development of an expanded (Muslim) public sphere, or as giving voice to those who have previously remained voiceless, notably Muslim women (Amir-Ebrahimi, 2004; Hermida, 2002). Notwithstanding this work's important contribution, there are several unanswered questions regarding the Muslim blogosphere. First and foremost, the question remains: what is the sociopolitical significance of Muslim blogosphere?

In addressing these questions, this chapter considers blogs as a novel cultural form, blogging as a cultural practice, and the blogosphere as a new collective and finds that they represent a unique opportunity to observe what (some) Muslims are actually doing and saying, and how they are involved in constructing both their identities as well as Islam more broadly. Notwithstanding the demographic specificity of bloggers, focusing on the Muslim blogosphere highlights Muslim voices as actively involved in the construction of their own identity/selfhood. Through looking at three themes emerging from the (Eurocentric) sociology of modernity—rationality, publicness, and individuality—this chapter shows that the Muslim blogosphere expands, and to a certain extent alters, set ideas on Western modernity and Muslim traditionalism. Concerning the political relevance of the blogosphere, it will be argued that this lies in its ongoing efforts to articulate demands for self-actualization and recognition of Muslim diverse identities with demands for redressing injustices and inequalities. Underpinning this argument is the idea that the Muslim blogosphere operates in a universalistic horizon that is based on ideas associated with modernity but which is expanded, negotiated, and amended in ways that seek to incorporate Muslim subjectivities.

Theories of Modernity and Islam

This section rather eclectically focuses on three theorists of modernity whose work has been very influential in understanding and configuring modernity: Max Weber, Jurgen Habermas, and Anthony Giddens. Weber's work has significantly contributed to understanding the role of rationality and rationalization in modernity. On the other hand, Habermas's work sought to involve processes and conditions of communication with rationalization and modernity, focusing on the process of critical publicity. Finally, Anthony Giddens sought to update theories of modernity leading to an emphasis on reflexivity and individualization. This section reviews these three theories, and their respective main claims, from the point of view of their implications for understanding Islam.

Weber, Rationality, and Islam

For Weber, the key to understanding the transition from tradition to modernity was rationalization. The main question underpinning much of his work concerns the processes by which societies rationalize. Although Weber used the term rationalization to refer to somewhat different processes, we can roughly define it as the systematic application of reason to solve problems and conduct one's life based on principles of efficiency rather than on emotions, relationships, superstitious beliefs, and so on (Weber, 1968). Rationality therefore is central to Weber's theorizing of modernity, and this section explores it further, along with its implications for Islam.

Weber understands rationality as a universal and species-based attribute and from this point of view Islam is understood as potentially rational but not yet rationalized. Weber's argument is a historical one: some societies (and social groups) differ in how, historically, they have applied principles of rationality. Specifically,

Weber referred to four types of rationality, formal, substantive, practical, and theoretical (Kalberg, 1980), which are encountered in most, if not all, historical societies and religions. However, the "West," or modernity for the two are often conflated, differs in its specific combination and application of all four types of rationality. To elaborate further, practical rationality refers to means-end types of activities that are aimed primarily to meet the individual's self-interests—this rationality is understood as instrumental, pragmatic reason. Theoretical rationality refers to the cognitive apprehension of reality and the world through processes of induction and deduction, attribution of causality, formation of symbolic meanings, and so on. This rationality can be further understood as primarily intellectual, and as such opposed to the emotions. Substantive rationality orders action neither on the basis of means-end self-interest nor on the basis of logical thought processes, but rather on the basis of a set of values endorsed by given societies and social groups. Weber outlined the ways in which this rationality leads to exclusions of other or opposing value-systems as "irrational," and from this point of view, we can understand substantive rationality as exclusionary. Finally, formal rationality refers to means-end types of action not on the basis of self-interest, as in practical rationality, but rather on the basis of universally applied laws, rules, and regulations—this can be understood as efficiency, the ratio of output to input.

Societies develop differently, primarily on the basis of adhering to different substantive rationalities. In ideal-typical terms, modernity is linked to a specific configuration of rationalities, one that revolves around pragmatic self-interest (instrumental practical rationality), scientific theoretical rationality, and bureaucratic formal rationality, mediated and coordinated through a value-rationalized substantive rationality, based on the values of the Enlightenment (autonomous and free individuals whose actions are given continuity through referring to ultimate values, see Weber, 1968; Kalberg, 1980:1176). Implied here is the view that Islam will, in time, conform to the requirements of the Western path to modernity through endorsing these types of rationality.

First and foremost, Weber's is a sociohistorical argument, and as such it needs to be revised on the basis of more recent historical developments on the one hand, and on the basis of observed social action on the other. This is where we can locate the significance of the blogosphere and blogging as a sociocultural action or practice. Sociohistorical changes and associated social practices may lead to substantial reconfigurations of rationality. There is in fact in Weber's work a clear recognition of the role of individual actions and social practices, which may give rise to new patterns of rationalization, new forms or ways of life, and ultimately new values and substantive rationalities. The first empirical question that we can pose vis-à-vis blogging as a social practice, therefore, is how does blogging, and in particular Muslim blogging, affect the ideal-typical rationalities discussed by Weber? But before addressing this question, we need to discuss the second pillar of modernity, and its relationship with Islam, that of publicness.

Habermas: The Public Sphere and Islam

Habermas's theory of the public sphere links modernity and rationalization through processes of communication. Habermas (1962/2006) uses the public sphere as a historical category, emerging under, and owing to, the convergence of certain social-cultural, political, and historical circumstances. The public sphere is seen not only as necessary for the functioning of modern democracies but also as a development closely associated with modernity and the rationalization of politics. The emergence of the public sphere meant that politics was no longer conducted under conditions of representative publicity, where monarchs paraded their authority, but rather under conditions of critical publicity, where political decisions are legitimated through rational-critical debate. When in the public sphere, citizens were expected to make "public use of their reason" (Habermas, 1962/2006:27), and it was through this that the public sphere acquired an overt political function.

The political function of the public sphere was to subject political decisions to rational-critical debate, thereby legitimating them. This proceeded in two ways: first, through making public political decisions, and the ways in which they were taken, and second, by subjecting such decisions to critical scrutiny. This is precisely the function of the news media: to print and generally publicize news about political decisions. Having publicized political decisions and facts, the public of citizens would then critically discuss them, forming a public opinion. This public opinion, in turn, had to be consensual as it was based on the application of reason, an attribute shared by all humanity: what is reasonable for one must also be reasonable for another. Thus, through the formation of public opinion, the public sphere contributed to the rationalization of political decisions. And through this application of reasoned argument, political decisions were legitimated, gradually removing authority from the state and giving it to the people, who ruled with their reason. For the public sphere to fulfill its function, access must be guaranteed to all, who must then enter it as private citizens, with no business, state, or other connections. Debate in the public sphere must take place on the basis of reasoned and informed argument rather than on the basis of interests and/or beliefs. The institutional context within which the political public sphere could operate was that of the nation-state, which comprised a more or less homogeneous community, sharing similar interests and concerns. Ultimately, the role of the public sphere is to reach a rational consensus.

In political terms, the requirement is that citizens enter the public sphere devoid of any interests, identities, and pre-conceived ideas: the public sphere must be secular and autonomous both from state and private interests. This means that Muslim people cannot participate in the public sphere as Muslim but as "human beings" in the abstract sense. But this abstract sense is in fact quite concrete, as it refers to a specific kind of person: an educated, secular, middle-class man, able and willing to use his reason in a specific manner ostensibly for the "common good."[1] In addition, if we accept that the ideal-typical context for the public sphere is an ethno-culturally homogeneous nation-state, then Muslim participants in Muslim minority countries are excluded not only because they are not allowed to participate *qua* Muslim but because they may often lack the sociocultural capital to do so. On the other hand, Muslim majority countries cannot be thought of as having proper public spheres insofar as they have not modernized sufficiently and place barriers to the development of free media and the open participation of all. At the same time, unless there is a proper and open public sphere in operation, governments in Muslim majority countries cannot be, and are not, held accountable for their decisions and actions. This has led to the formulation of a paradoxical relationship between the public sphere and Muslim majority countries: a public sphere cannot operate under conditions of insufficient modernization and rationalization, while such modernization and rationalization cannot come about without the contribution of a vibrant and open public sphere.

Notwithstanding these exclusions, the theory of the public sphere offers the potential to groups to expand and modify it in and through its own principles—that is, it allows people to make use of critical rational arguments to make demands or claims, because of its expanding universalism and the firm belief that reason and rationality underpin humanity as a species. At the same time, in the actual application of their reason, people may modify or expand the public sphere or apply their communicative actions in unpredictable ways. Thus, despite the many and often justified criticisms addressed to the theory of public sphere and communicative action, it offers a dynamic view of modernity which allows for change and expansion. From this we can obtain the second set of empirical questions: what communicative actions are undertaken by Muslims blogs vis-à-vis the question of publicness? How is publicness itself modified in the Muslim blogosphere and through blogging? What is the relationship of the blogosphere to the public sphere?

Individualization: Giddens, Modernity, and Self-Identity. The final theoretical thread to be followed here concerns the relationship between the self or subjectivity and modernity. As with the above two theories, Giddens's perspective is a historically nuanced one, understanding modernity as having introduced "an elemental dynamism into human affairs" (Giddens, 1991:32). Giddens looks at late modernity, focusing on questions of risk and uncertainty. Rationalization has not fulfilled its promise to rid the world from chaos, injustice, and domination; this, along with the loss of traditional anchors, such as religion, custom, and the ties of kinship, led to a era in which risk, insecurity, and anxiety predominate. Giddens understands modernity as a risk society, in which none of our activities can follow a predestined course, and are always subjected to contingent events (Giddens, 1991:28; cf. Beck, 1992). In managing such contingency and risk, modernity introduces the dimension of reflexivity: "the regularized use of knowledge about circumstances of social life" (Giddens, 1991:20). Giddens argues that late modernity leads to an increased relevance of two opposite dimensions: first, globalization, understood as the embodiment of the universalizing tendencies of modernity; second, locality, or the local engagements with the processes of globalization. Giddens locates the relevance of the self and identity at the level of the local, which he understands as dialectically related to the global.

For Giddens, the self is no longer anchored in space and time and in established social and familial networks. Moreover, it needs to deal with abstract, expert, and impersonal systems in ways that make sense and preserve the self's integrity. Giddens argues that in order to do so, the self must develop reflexively, that is, it must constantly refer back to itself, feeding into it knowledge accumulated in other spheres of life. Rather than looking at this type of self as a kind of loss of certainty and security, Giddens understands it in more positive terms. The loss of traditional anchors opens up not only a world of risk but also of possibility. The trajectory of the self is no longer given from birth through adulthood to old age and death, but rather people can creatively intervene in their own lives and live them in ways they choose. As Giddens puts it, the self becomes a reflexive, ongoing project (1991).

In political terms, Giddens argues that these developments lead to the formulation of a life politics, which stands in contrast to a politics of emancipation. While emancipatory politics seeks to free social life from the shackles of tradition and custom, life politics views politics as linked to freedom of choice and generative or transformative power (the power to change things). In addition, while emancipatory politics is concerned with the elimination of exploitation and the equal redistribution of resources, life politics is concerned with the creation of forms of life which promote self-actualization. Finally, where emancipatory politics follows the ethics of justice, equality and participation, life politics develops ethics concerning the question "how should we live?" (Giddens, 1991:215). In other words, Giddens considers life politics as the politics of late modernity and views it as concerned primarily with questions of self-actualization in post-traditional, globalized settings.

Giddens's theoretical efforts exclude Islam no less than the earlier two theories. First, Muslim subjectivities are excluded because they cannot be understood as "projects" built on the basis of individual choices. Insofar as Muslim subjectivities are obtained through an engagement with religious beliefs and texts, they can only be seen as ascriptions rather than constructions. Second, the project of the self has emerged as a response to an age of risk, doubt, and uncertainty; however, religious and transcendental beliefs and teachings are incompatible with doubts, ambiguities, and uncertainties, or at least entail a set of prescriptive behaviors for dealing with these. The concepts of responsibility and choice linked to the "project of the self" are in Muslim identities-as-ascriptions conspicuously absent, thereby positioning them as devoid of any differentiation and accountability. In other words, religious identities, within the context of Giddens's theory, can be understood only as traditional identities. Giddens (1991:207) in fact understands the affirmation of religious identities as a reaction to the inability of modernity to effectively provide a new ethico-moral framework for identity construction—in Giddens's version, this return of religion will no longer be necessary if life politics prevails. Nevertheless, Islam remains excluded whether seen as traditional or as an inappropriate response to the current phase of modernity. In addition, although life politics represents in some ways the extension of politics in other domains, its demands ring hollow in the face of the continuous exploitation, oppression, and discrimination experienced by Muslims in Western Muslim minority countries as well as in war-torn, and/or illiberal Muslim majority countries.

From the above discussion, we can discern the increased relevance of the self and subjectivity in late modernity. Looking at Muslim blogging as a cultural practice, we can then pose the following empirical questions vis-à-vis the issue of the self and subjectivity: how does Muslim blogging and the emerging blogging subjectivities relate to the "project of the self"? What type of politics is the Muslim blogosphere involved in?

Cultural Practices: Blogging and Islam

The relevance of the blogosphere is three-fold. First, as we have seen in the above theoretical discussion, although social theory acknowledges the role played by sociocultural practices, it has overlooked Muslim social practices and lived experiences focusing mainly on scripture and social structures. From this point of view, the blogosphere might be thought of as an aspect of lived experience and a sociocultural practice that (re)constructs and (re)interprets established understandings. Second, we may consider the blog as a novel cultural form and practice that significantly differs from other such forms in that it popularizes and positively encourages self-expression and publishing and is accessible to all (those with Internet connections). In other words, blogging is different from working because it is primarily undertaken during leisure; it is different from watching TV or reading because of the type of activity involved, different from socializing because most blog-gers do not actually meet their readers in person, different from e-mailing because it's public, and so forth. As a unique and novel sociocultural activity blogging requires understanding and contextualization. Finally, looking at the Muslim blo-gosphere specifically allows us to rethink the assumptions of social theory when it comes to Islam; such a move is premised on the importance of sociocultural prac-tices in shaping Islam, and other sociocultural identities, thereby recognizing the inherent and irreducible dynamism of any identity.

The discussion here is meant to be an initial exploration of these three ques-tions rather than an exhaustive study. The empirical material is based on reading and following about 25 blogs for about nine months, from January to September 2007. All these blogs are written by bloggers explicitly identifying themselves as Muslim or who refer to their Muslim origins regardless of whether they are practicing or not. The label Muslim is here used in a very broad way that includes both those choosing it because of their religious beliefs and those who "inherit" it. In addition, the blogs focused on themes that were linked to Islam or the experiences of being Muslim. Thus, a blog about deciding to wear the hijab, for instance, was included, while a blog by a Muslim blogger on the topic of computer networks was not includ-ed. Finally, the blogs followed here were all in English. From this point of view, this analysis cannot and does not claim representativeness. Rather, it is acknowledged that the diversity and dynamism of Muslim identities and experiences prevent any broad generalizations to be made. At the same time, however, observing sociocul-tural practices might offer us important insights in the ways in which social theo-retical understandings of modernity are in practice negotiated, rejected, or otherwise altered.

Following the theoretical discussion, we can sum up the main three questions emerging as follows: What is the relationship between Muslim blogs and rational-ity? How is publicness reconfigured in Muslim blogs? And what type of subjectiv-ity emerges in Muslim blogs and what is its political relevance?

Muslim Blogs and Rationality

The discussion on Weber suggests at least four understandings of rationality. This section follows these and sees what Muslim blogs do with/to them. These include theoretical rationality, substantive rationality, practical rationality, and formal ratio-nality. All these generate oppositions and exclusions that are negotiated, resisted, or expanded. However given the added significance of theoretical and substantive rationalities, these are the ones discussed here.

Theoretical rationality refers to a cognitive apprehension of the world, an intellectual grasping of its main structures, principles, and functions, and the reasons or causes behind them. In these terms, this rationality can be understood as the opposite of emotionality. In Muslim blogs, we find both a theoretical, cognitive rationality that seeks to describe and explain diverse events and phenomena through reasoned arguments as well as clear expressions of emotionality. The concurrent existence of both in blogs disputes the claim of theoretical rationality as the only or main means of apprehending the world. The very title of As'ad Abu Khalil's blog, the Angry Arab News Service, points to an emotion, anger, mobilized as a response to events in the Middle East and elsewhere. But the Angry Arab also expresses other emotions, especially grief and sadness, and often makes use of humor in his posts. For instance, in his blog's home page, the Angry Arab has posted photographs of two obviously distressed girls screaming, followed by the caption "Iraqis react to Bush's 'liberation' of their country." He also has a feature that calls for the canonization of Mother Theresa, asking readers to make up miracles. In other posts, he mixes anger with irony and sarcasm.

At the same time, however, his blog contains sophisticated analyses of events, either written by him, or else linked to the site. Abu Khalil is a Lebanese political scientist, teaching at California State University, so most of his analyses are informed, well researched, and elegantly written. For instance, he provides a link to a chapter he has written on the prolonged Lebanese civil war, although he also writes less academic commentaries that he then posts on his blog or publishes elsewhere, for example, in the press, or on Salon.com, and links to his blog. Some of his commentaries are written in Arabic. In addition, he provides links to analyses found elsewhere: for instance, he provides a link to a *Washington Post* article titled "Lessons in Forced Democracy." Abu Khalil is not the only blogger mixing emotional responses with serious argument and analysis. For instance, Yahya Birt's blog is more focused on serious analysis, and contains several links to academic works, but he also has a category on humor listed in his blog. Although Yahya Birt is also an academic and a leading figure in the British antiwar movement, other, less academically inclined bloggers also provide their own analyses, often mixing emotions with arguments—the blog of Suspect Paki (100% Londoner, 100% Muslim. Deal with It) has a post that critically analyzes an article by *The Guardian* on Israeli bombardment of Gaza, barely disguising the author's anger at the continued injustice suffered by Palestinians.

This concurrent existence of both intellectual, reasoned analyses and arguments with emotional expressions and responses points to an evident attempt to apprehend the world not only cognitively, in its abstract dimensions, but also emotionally, in its concrete effects as experienced and felt by human beings. From this point of view, we may argue that the abstract theoretical rationality is complemented by an emotional rationality, which combines the cognitions with the emotions. Although such a combination occurs in other spaces too (cf. Nussbaum, 2001), we must insist that blogging by Muslims, long stripped of their cognitive abilities for original analyses through Orientalist assumptions, not only restores Muslim rationality, but also extends it to encompass the emotions. It is not a hysterical irrationality that one finds in Muslim blogs but thoughtful analyses that justify and require emotional responses.

Substantive rationality is understood by Weber as the kind of rationality linked to cultural values associated with specific communities. It can be therefore be understood as excluding and even actively opposing other value-rationalities. Weber's idea was to develop a substantive rationality not only of tolerance, but rather one based on reason and hence being common to all. However, in the Muslim blogosphere there are several examples of active rejection of other value-rationalities, albeit ranging widely and including Islamism, Christian fundamentalism, Sunni Islam, Shi'ite Islam, Judaism, liberal atheism, capitalism consumerism, and neoconservative ideology, to name but a few. Posts here can be outright polemical, clearly showing what Weber described as the inability of one value—rationality—to accept another. Animosity is expressed not only in blog posts but also often in comments and replies. Certainly common to all blogs was an opposition to the ideology and values behind the war in Iraq—expressed in various ways as anti-Americanism, anti-Bushism, neoconservatism, or sometimes as anti-Semitism. Substantive rationality can therefore lead to exclusion of other value-rationalities, or at least to a degree of polarization and separation between them.

However, common in many blogs was an attempt to describe, explain, and to make readers understand Islam and Muslim values. Very often bloggers cite verses from the Qur'an, or poetry from well-known Muslim poets, to illustrate of their feelings, values, and experiences of being Muslim. In addition, they clearly explain Islamic festivals, fasts, and ceremonies, such as the Hajj, Ramadan, and so on. Some of these posts are clearly addressing non-Muslim readers whereas others address community members with advice. For instance, Cool Guy Muslim's blog has a post giving tips for preparing for Ramadan. In another post in the same blog, Cool Guy Muslim is telling parables about Muslim norms of behavior. While the posts with the verses and the description of important Muslim occasions might be interpreted as contributing to understanding and reconciliation, the posts on advice and tips might be seen as strengthening shared values, while also to some extent policing and safeguarding Muslim values. Another relevant feature here concerns the comments and links provided by the blogs. These open up a dialogue, or at least allow for the possibility of a dialogue between different value-rationalities and also within Muslim substantive rationality. Thus, anything said, claimed, or argued might encounter opposition, disputation, and even rejection. This feature both supports and undermines understanding and reconciliation as well as any attempt to reduce and control the diversity of Muslim beliefs and practices. Thus, Muslim bloggers both reproduce the exclusions immanent in having a value-rationality, as well as open up their own value-rationality to others' understanding, and thereby to change, while at the same time seeking to control and safeguard it.

This analysis shows that rationality is not undermined but rather endorsed by Muslim bloggers. In the end, what we can see here is a tempered, extended rationality, which is widened and complemented, and in this manner, possibly contributing to an alternative rationalization.

Muslim Blogs and Publicness

As with rationality, publicness takes many forms. First, it is understood as inclusiveness, openness, and dialogue: everyone can join in and anything can be discussed, provided that it pertains to the domain understood as public. In oppositional terms, public is understood as the opposite of private. Second, publicness involves criticism: critical arguments must be made. This, in turn, is linked to the issue of consensus: insofar as publicness entails inclusive, critical, but reasoned dialogue, it is expected to end in consensus, as the most reasonable argument will prevail. This section examines the ways in which these understandings are reconfigured in Muslim blogs.

Publicness as openness and inclusiveness is at the heart of the public sphere theory. Allowing any topic for discussion and allowing people to discuss topics regardless of who they are are two of the main conditions for the proper functioning of the public sphere, or for communicative action to take place. First, inclusiveness in a sense goes with the territory: blogging as a medium is accessible to all those with Internet access, as new interfaces make it very user friendly. Equally, the ability to post comments on blog entries forms another dimension of this inclusiveness. The comments left on the Muslim blogs under study reveal that readers include both Muslims and non-Muslims of diverse opinions and backgrounds. In terms of openness, the Muslim blogs studied here certainly cover a very wide range of thematic categories, including religion, politics, sexuality, personal life, and so on. For instance, KABOBfest, a collaborative blog by Arab-Americans has some 91 categories, ranging from activism to "zionuts," covering controversial themes such as cartoons, gay rights, and Palestine, alongside more expected topics, such as Islam and the war on terror ("terrorgasm"), next to more cultural themes, such as hip-hop, culture, and books. Other blogs, such as the Bradford Muslim, do not have categories but list their previous posts, which again cover a wide range of issues. Of interest here is the almost seamless merging of the political/public themes with the more personal/private ones.

Indeed, for Habermas there must be a separation between deliberation on topics that pertain to public life, which require discussion and eventually agreement, and those that concern the private domain, which are more matters of preference, individual choice, and do not require any form of consensus. Nevertheless, Muslim experiences show that decisions typically classifiable as private in the context of European modernity, such as to convert, to worship or to wear a hijab, or to be sexual beings, become issues for public discussion. Bloggers then write blog entries that pose such issues for discussion, while in some cases the whole blog is dedicated to discussing such experiences—for instance the blog by Scarf Ace, in which the author documents her decision to wear the hijab in the context of working and living in the United Sates. In most of her blog entries the author wants to make clear the reasons underlying her decision, the links to scripture, the variations in opinion regarding codes of dressing and piety—in short, to invite debate and discussion on a matter that would, under different circumstances, be a personal choice. As the feminist critique of the public sphere has shown, the lines between private and public are continuously shifting, and perhaps blogs might be thought as contributing to such shifts. However, this development is not necessarily a positive one: underlying Scarf Ace's attempts to publicly explain her decision to wear a hijab is a climate of persecution felt by many Muslim women in secular countries. And from another point of view, some blogs problematize and bring to the fore issues that have been resolved in liberal societies—for instance, the blog "Eye on Gay Muslims" aims to "support" gay Muslims in reverting to heterosexuality, as homosexuality is seen as a major sin. It seems therefore that Muslim blogs foreground the ambivalence of publicness: on the one hand, openness to all themes allows for revision, discussion, and perhaps more equitable and just solutions to ongoing problems; on the other, it may lead to a negative kind of revisionism, and in some instances outright regression. It's not that Muslim blogs are causing this—rather, this seems to be an inherent ambiguity in the heart of publicity. Nevertheless, because of the shifting between lines of public and private precipitated to an extent by blogs, this ambiguity is highlighted through Muslim blogs.

In addition, Muslim blogs are actively involved in both dialogue and critique. To begin with, the blog features of comments, tagging, trackbacks, and permalinks certainly encourage an exchange between bloggers and readers, while they can also move the discussion across blogs. Although blogs differ in terms of whether they allow readers to comment freely or moderate comments, they all allow comments. Though exchange is part of blogging, the extent to which it constitutes dialogue is questionable. Although some comments are thoughtful and interesting, most appear to include almost random thoughts of readers, and it is rare to find the bloggers themselves responding to comments. Moreover, where comments are not moderated, the exchanges may not even be within the bounds of civility. This is very often the case in the Angry Arab and Radical Muslim blogs, which although attracting several comments, tend to be abusive and outright racist. Muslim blogging as a cultural practice therefore initiates dialogue, but this is not a guarantee that dialogue will indeed take place. Because in this sense blogging is but the beginning of a dialogue, we cannot expect any form of agreement or consensus to emerge. On the contrary, what is very often encountered in Muslim blogs is outright disagreement, coming from both other Muslims, and non-Muslims. From this point of view, the work of Muslim blogs is not to conduct dialogue or deliberations on a topic, but more to begin or to kick-start discussion on issues and topics of concern, even if they are controversial and divisive.

In these terms, it is highly relevant that all Muslim blogs under study were very critical. Critiques ranged in terms of their contents, but by far the most popular issues for critique included the war on terror, Iraq, and Palestine. Bloggers differ as to the extent of their polemicism as well as in terms of the creativity with which they oppose the war. Some use words and links to other sites of similar views, while others have pictures, call for petitions, counter with deaths tolls, and so on. In addition, there is a range of styles associated with critique, from parody to informed criticism. For instance, the Islamicist blog parodies Islamism, spoofing Ed Husain's book *The Islamist* (2007). In its entries, it satirizes Islamist speak and ideas, from the Caliphate to women's modesty, and through this parody it creates space for a critique of some positions developed among some Muslims. In contrast to the light style of the Islamicist blog, other blogs provide serious and informed analysis in order to recognize and criticize injustices. Such blogs include Yahya Birt's blog, and the Cutting Edge blog, alongside the Angry Arab blog, and KABOBfest: what they have in common is critical, informed, incisive commentary on the war on terror, Iraq and Palestine. This is more than the stating of opinion: rather it is the provision of sustained and persuasive argument, which, moreover, is not subjected to any kind of censorship, either direct or indirect. As such, these blogs widen the debate on contested issues, and complement, expand, and some times effectively counter, the mainstream media coverage of such issues. In this manner, Muslim blogs fulfill an important function of the public sphere.

Muslim blogs appear to reconfigure the public sphere along four significant lines: first, they expand and blur the lines separating the private from the public. This is not only by means of publicly writing on matters considered private—this is certainly something very often encountered in all types of blogs. Rather, Muslim blogs actively question the assignment or delegation of some issues as private, by showing the ways in which they have become politically relevant—in the sense that they concern matters of distribution of power, material, and/or symbolic. Second, Muslim blogs foreground some of the inherent ambiguities of the public sphere: being necessarily open, it must allow for discussion of matters that are for some already resolved. Third, rather than conducting dialogue, they invite dialogue that may take place elsewhere, or indeed may not take place at all. Similarly, the critical edge of Muslim blogs widens the public sphere, as people now do not have to rely exclusively on mainstream media. At the same time there are no guarantees that these critiques will be read, taken in, and acted upon. In these terms, the type of public sphere associated with the Muslim blogosphere appears to be a truncated one, in which only some of the public sphere's functions are extended and amplified. From this point of view, the political function of the public sphere as the space/means by which public opinion is formed, subsequently legitimizing political decisions, cannot be guaranteed, at least insofar as dialogue and deliberation do not take place. The question of the political function of the Muslim blogosphere therefore remains open.

Muslim Blogs, Identity, and Life Politics

In his discussion of modernity and self-identity, Giddens first understands identity as separateness, that which makes us unique human beings, and secondly, he argues that in late modernity, the loss of traditional anchors means that self-identity is a matter of personal construction rather than ascription, and that having a self means being in charge of an ongoing project. Finally, he argued that the project of the self leads to a new politics, a life politics concerned more with self-actualization than with emancipation. This section examines the type of Muslim identity and subjectivity emerging from blogging, and the kind of politics it can be linked to.

As argued earlier, Muslim identities are considered to be more or less uniform and Muslim people appear interchangeable with each other, rather than unique individuals. This is a common finding in analyses of Islamophobia, which positions all Muslims as the same (Runnymede Trust, 1997). However, blogging acts in ways that effectively establish the separateness and uniqueness of the blogger, first by naming and authoring a blog, and second through having a unique URL address. Thus, Muslim bloggers select specific names for their blogs, and assume authorial responsibility for what they write and post. The use of the pronouns "I" and "me" is especially common in blogs, and it denotes on the one hand precisely this subjective element of the blog, and on the other the claiming of the blog as part of the blogger's identity (Amir-Ebrahimi, 2004; Reed, 2005). Significantly, the Muslim blogs under study, as well as other Muslim blogs glimpsed through links and blogrolls, appear to choose names that somehow relate to being Muslim—Radical Muslim (Politically charged writings of Muslim in Britain), Bradford Muslim (Notes on life in a multicultural city), Suspect Paki (100% Londoner, 100% Muslim. Deal with it.), iMuslim (British blogger, Desi Dreamer, Mad Muslimah), and so on, all denote a Muslim identity on the one hand, and another more particular and subjective viewpoint on the other hand, which renders these blogs and their perspectives unique. This Muslim identity changes and shifts as a function of the elements that designate the blogger as unique, and through the blogger's ongoing reflection and engagement with the world and the self. The uniqueness and separateness of Muslim bloggers is complemented by movements that denote an ongoing collaboration with others and a certain degree of belongingness. Thus, blogrolls and links to "friends," "comrades," "brothers/sisters," or even enemies, as in Radical Muslim's "fatwaworthy" sites, all show that each blog might be unique, but does not exist in isolation. Rather, they operate in a context that they help define, and which, in turn, helps define them. Although separate and unique thereby denoting an individualized identity in place, blogs point to something else as well: a form of collaboration, of coexistence that in some ways casts doubts on a model of identity as a self-construction.

More specifically, the project of self-identity refers to the ways in which individuals, in conditions of late modernity, pick and choose who they want to be mainly in terms of lifestyle and consumption choices. This implies that they somehow operate in relative isolation from each other, almost cut off from their social experiences in given sociocultural contexts. The Muslim blogs' sociality, in contrast, shows how blogging combines identity-in-the-making with connection with others, who also help define this identity. Thus, for a blogger to choose and list their friends and their enemies, to write about who and what they like or not, show that they mutually define each other, within the context in which they operate—in this case the Muslim blogosphere. On the one hand, therefore, the selectivity of the blogs, the fact that the blogger has discretion over most parts of the blog, shows that indeed to some extent they define and construct their own identity, while on the other the mobilization of other blogs or blog entries in doing so, denotes the collaborative or social elements of this construction. For Muslim blogs, this points to the broad commonality of beliefs shared with others, and the uniqueness of experiences, viewpoints, and opinions, shaped within this context but uniquely claimed by the individual blogger.

Moreover, both the unique and collaborative or shared aspects of the Muslim identity emerging through blogging are not obtained through references to lifestyle and consumer choices but rather through writing, reflection, argument, and justification. From this point of view, blogging appears closer to the politics of self-actualization posited by Giddens, as it concerns "debates and contestations deriving from the reflexive project of the self" (Giddens, 1991:215), particularly that of being Muslim in the current political climate. The political significance of [Muslim] blogging therefore lies in the ways in which it allows bloggers to redefine and actualize their identities as Muslims, on the one hand, and as unique individuals on the other. However, this is only half the story: injustices, negative experiences, exile, persecution, poverty, outright racism, and discrimination are some of the experiences written about, and often shared by Muslim bloggers. In writing, criticizing, and arguing about and against these injustices, Muslim blogs actively participate in a politics of emancipation that seeks to foreground injustice, exploitation, and inequality, protest about them and ultimately change them. The efficacy of this politics is of course under question. Our discussion of publicity showed that neither agreement nor consensus exists in the Muslim blogosphere. The uniqueness of individual bloggers and the associated diversity within the Muslim blogosphere show that there are no easy solutions to issues of injustice and inequality. Nevertheless, the political role of the Muslim blogosphere might not be to provide solutions but to foreground problems, not to formulate and lead opinions but to ensure that ongoing injustices are not forgotten. Perhaps then, the unique contribution of the Muslim blogosphere is to articulate a life politics of actualization with an ongoing search for a politics of emancipation.

Conclusion

In bringing together the rather abstract theorizations of modernity and the grounded cultural practice of blogging, this chapter sought to demonstrate how modernity is reconfigured through sociocultural practices. While social theory either by commission or by omission excluded Muslims from a modernity revolving around rationality, publicness, and individuality, we tried to show how a novel form of sociocultural practice may be associated to new understandings of modernity.

Blogging was therefore understood primarily as a sociocultural practice, incorporating at the same lived experience and reflection upon this experience, and as such opening up a new window of understanding. Indeed, an empirically based discussion showed that Muslim blogging expands rationality, complements it with emotions, and turns more toward understanding rather than solution-oriented instrumentality. Moreover, Muslim blogging extends the dividing line between the public and the private, opens up the debate, and eschews consensus in favor of active disagreement and critique. All these, however, have ambiguous repercussions and debatable political functions. Finally, Muslim blogging reconfigures identity as both shared-ness and uniqueness and initiates a combination of a politics of identity actualization with a politics of emancipation.

What does all this mean for modernity? Rather than undermining any of what we here designated as its three pillars, we see that Muslim blogging extends, complements, and combines them. This leads us to argue that modernity (rationality, publicness, individualization) becomes a universalistic horizon across which different local scenarios/interpretations are enacted, with the possibility of subsequently feeding into this framework, ultimately ensuring its dynamism. Blogging as a cultural practice must be seen as operating within this universalistic horizon, within a context in which certain claims are made and considered applicable to all. But this universality must not be seen as already predetermined and set up in stone. It is a malleable universality that is dynamic to the extent that it allows itself to shift and change according to new ideas, practices, and so on. The point here is that unless we take Muslim blogging seriously as contributing to this universalistic horizon against which it first emerged, its positive contributions will be lost: the understanding it seeks will be lost, the debate it attempts to initiate will dissipate and become increasingly polarized, and its effort to combine life with emancipatory politics will fail. This in a sense sets up the future agenda for the [Muslim] blogosphere: to draw the ways in which blogging practices can reenter or feed back into both the formal, institutional aspects of life, and into the everyday politics of coexistence.

Note

1.   See for more details the feminist critique of the public sphere: Fraser, 1992.

References

Alexanian, J. (2006). Publicly intimate online: Iranian web logs in Southern California. *Comparative Studies of South Asia, Africa, and the Middle East, 26*(1), 134–145.

Amir-Ebrahimi, M. (2004). Performance in everyday life and the rediscovery of the "self" in Iranian weblogs. *Bad Jens Iranian Feminist Newsletter*. Retrieved from http://www.badjens.com/rediscovery.html on September 10, 2007.

Beck, U. (1992). *Risk society: Towards a new modernity*. London: Sage.

Burns, A., & Jacobs, J. (Eds.). (2006). *Uses of blogs*. New York: Peter Lang.

Fraser, N. (1992). Rethinking the public sphere: A contribution to the critique of actually existing democracy. In C. Calhoun (Ed.), *Habermas and the public sphere* (pp. 109–142). Cambridge, MA: MIT Press.

Giddens, A. (1991). *Modernity and self-identity*. Cambridge, MA: Polity Press.

Habermas, J. (1962/2006). *The structural transformation of the public sphere*. Cambridge, MA: Polity Press.

Hermida, A. (2002). *Web gives voice to Iranian women*. Retrieved from BBC News Online Web site:http://news.bbc.co.uk/1/hi/sci/tech/2044802.stm on September 10, 2007.

Husain, E. (2007). *The Islamist*. London: Penguin Press.

Kalberg, S. (1980). Max Weber's types of rationality: Cornerstones for the analysis of rationalization processes in history. *American Journal of Sociology, 85*(5), 1145–1179.

Nussbaum, M. (2001). *Upheavals of thought: The intelligence of the emotions*. Cambridge, MA: Cambridge University Press.

Reed, A. (2005). "My blog is me": Texts and persons in UK online journal culture (and anthropology). *Ethnos, 70*(2), 220–242.

Runnymede Trust (1997). *Islamophobia: A challenge for us all*. London: Runnymede Trust.

Tremayne, M. (Ed.). (2007). *Blogging, citizenship and the future of media*. New York: Routledge.

Weber, M. (1968). *Economy and society: An outline of interpretive sociology* (E. Fischoff et al., Trans.). New York: Bedminster Press.

List of Blogs under Analysis

1. Bradford Muslim, http://bradfordmuslim.blogspot.com/
2. Cool Guy Muslim's, blog, http://mustaqeem.wordpress.com/
3. Eye on Gay Muslims, http://gaymuslims.wordpress.com/
4. Happy Arab, http://happyarab.blogspot.com/
5. Hijabi Apprentice, http://hijabiapprentice.blogspot.com/
6. Hypocalypse, http://www.hypocalypse.com/home/
7. iMuslim, http://imuslim.wordpress.com/
8. Indigo Jo blogs, http://www.blogistan.co.uk/blog/
9. Iraqi Konfused Kid, http://ejectiraqikkk.blogspot.com/
10. Iraqi Mojo, http://iraqimojo.blogspot.com/
11. KABOBfest, http://www.kabobfest.com/
12. Mustaqim, http://flyingimam.blogspot.com/
13. Radical Muslim, http://radicalmuslim.blogsome.com/
14. Raising Yusuf, http://www.a-mother-from-gaza.blogspot.com/
15. Rolled-Up Trousers, http://www.osamasaeed.org/
16. Scarf Ace, http://scarfacewearingaheadscarfinamerica.blogspot.com/
17. Serious Golmal, http://golmal.pickledpolitics.com/
18. Spirit 21, http://www.spirit21.co.uk/
19. Sugar Cubes, http://thesugarcubes.net/
20. Suspect Paki, http://www.suspectpaki.com/
21. The Angry Arab News Service, http://angryarab.blogspot.com/
22. The Cutting Edge, http://www.nafeez.blogspot.com/
23. The Islamicist's Weblog, http://theislamicist.wordpress.com/
24. The Neurocentric, http://neurocentric.blogspot.com/
25. Yahya Birt, http://www.yahyabirt.com/

CHAPTER FOUR

Bridges OR Breaches?

Thoughts on How People Use Blogs in China

KIM DE VRIES

Blogs have garnered increasing attention as researchers observe the important role blogs play in identity formation and expression for their writers and their increasing intersection with mainstream media. In particular, bloggers may use the perceived safer space of the blogosphere to experiment with their own identities and with new ways of defining themselves for an imagined international audience. If they become popular, these writers may be confronted with responses hitherto reserved for professional news outlets, in particular challenges to their authenticity or character. This chapter explores the way bloggers writing in English from China interact with readers. For example, these bloggers must decide what aspects of their own identity to represent and think about what it means for them to be blogging in English from China. Expression of ethnic or national identity in blogs is of interest in this time of increasing globalization because the blogosphere offers porous "national borders" around spaces in which participants and visitors can test different ways of defining a new kind of cosmopolitan identity.

These blogs create opportunities for communication not only with non-Chinese who are interested in Chinese culture, but also within a local Chinese community, and a larger Chinese or Asian diasporic community. Further, research by Lisa Nakamura and others suggests that instead of abandoning ethnicity and other markers of identity as was originally predicted (Turkle, 1995), bloggers and others who participate in online communities deliberately identify themselves in these terms of ethnicity and culture and take great interest in other members'/writers' cultures and ethnicities as well. In particular, by examining a range of English blogs written from China, this chapter considers how a distinct ethnic or cultural identity is expressed in light of an audience perceived as comprising both local and international readers and how communication practices reflect this perception when bloggers and their readers interact. Further, examination of blogs written from China in English suggests that contrary to concern that they serve primarily as another conduit for Western cultural imperialism (McChesney 2001; Virilio, 1995), the blogs have helped to define a robust national identity and undermine stereotypical images of China.[1] This chapter looks at five blogs that have all been online since 2002 and treats author entries and statements in interviews and lectures as primary source material.

The blogs discussed in this chapter have been chosen for several reasons: they each began in 2002 and number among the earliest blogs written from China; they are written in English, which not only facilitates this research, but more importantly allows certain assumptions about the blogger and the potential audience; and finally, the writers of these blogs have all become prominent in the Chinese blogosphere, as evidenced both by their being interviewed, quoted, and invited to speak at a variety of events, and by the way some of them have themselves organized important groups, events, or Web sites. I do not argue that these, or any five blogs, can represent the Chinese blogosphere in all of its growing complexity. Because these blogs have become so prominent and are so often linked to by others and the authors have gained such wide repute, however, we can take them to represent what many members of the blogosphere see as excellent blogs. Finally, all of these blogs represent "bridgeblogs," meaning that they are linked to by bloggers in at least two other nations that don't share a border or language (Zuckerman, 2005:9). Before looking at the way these bloggers write and speak about blogging in China, and as part of their lives, a brief history of the Chinese blogosphere will be helpful.

Though the earliest blogs in the United States mainly collected links and news stories with brief comments from the author, most histories of the blogosphere agree that in 1999 when Blogger.com's easy user interface allowed less technologically skilled users to easily post entries, the blogosphere exploded and its content shifted to include many blogs that work as online journals or diaries (Blood, 2000; Jensen, 2005; Mortensen & Walker, 2002; *Pew Report*, July 19, 2006). These personal diary-style blogs focus on individual thoughts and feelings and grew out of a decidedly Western romantic view of writing in which each author's first audience is him- or herself, and primary responsibility is to some interior truth, rather than to a larger community. By the time blogging spread to Asia, the "romantic" aspect of blogging was in full bloom. This focus on the personal and individual arguably makes blogs culturally alien to societies we tend to define as more collectivist in orientation, such as China, and some may fear explosion of blogging there represents yet another instance of cultural imperialism in which Western technology replaces or preempts local development.[2] However, some of the bloggers themselves, including those discussed here, claim the blogs can lead to greater understanding between China and the rest of the world and seem to ignore the question of cultural imperialism entirely.

Some reports on Chinese bloggers might seem to support the former view. In late 2005, a *South China Morning Post* reporter covered the first CNBloggerCon, and wrote that "the fact that so many are now broadcasting their individual tastes and preferences in music, fashion or food online hints at a deeper cultural shift in Chinese society and a new taste for individualism and openness over conformism and hierarchy" (Wong, 2005). There is no disputing that Chinese culture is changing, but rather than driving the changing, other evidence suggests that the popularity of blogging reflects changes already underway in China. We can more accurately say that blogging has accelerated some of those changes, but the rapid growth of the Chinese blogosphere would not have occurred if blogging did not appeal to already existing desires among Chinese netizens (MacKinnon, 2007; Mao, 2007).

The concern about cultural imperialism is certainly legitimate, given the history of relations between Western nations and China, but in this case it is based on several erroneous assumptions: that blogging represents a new encounter with Western values of written expression; that Chinese bloggers have not had a hand in shaping the blogosphere but instead can only conform to existing standards and expectations; and finally on a paternalistic view of Chinese people as unable to choose how or whether to use blogs. First, blogging is sometimes represented as valorizing a self-centered or self-absorbed approach to the Web and this might be seen as an extreme of Western individualism now infecting the Chinese culture. But blogs are far from China's first encounter with a Western focus on self-exploration through writing (Li 1995). Since 1980, China has been importing native English speakers as English teachers in rapidly growing numbers, and since the late 1990s, many western schools and academies have opened Chinese branches (Turner & Acker 2002). Both developments have led to writing instruction in China coming to follow a Western approach that focuses on the personal essay and on highly individualistic self-expression (Li; Turner & Acker). In fact, some of the earliest bloggers in China started blogging as a way to practice their English (Leylop, 2006).

Next, if blogging started to take off in the United States in 1999 (Blood, 2007), Chinese bloggers were not far behind. Though blogging from China is generally dated from 2002 (Qiang, 2004), people in China were already online as part of a global diasporic community. For example, Chinese professionals and students in particular were posting to online communities and LIST SERVS from

America, Britain, and Canada, while Western students of ethnically Chinese origin were writing from China as a way to stay in touch with friends and to help them negotiate their identities in a new setting. Both groups met in communities like the Chinese Community Forum, which existed from 1993 to 2005 and GenerationRice, which operated from 1999 to 2004 (Lee & Wong, 2003). Looking back over the archives of these kinds of sites, which were run more like newsletters, sometimes with attached bulletin boards, we start to see a drop in participation in 2003–2004, just as blogs became widely popular in China and as technological advances made blogging in Chinese easier (Mao, *Meta*, November 2004, Keynote, 2005). We can draw two conclusions from this history: first, that given the choice, many Chinese netizens preferred blogging to writing for a newsletter, and hence have a more individualistic approach. Second, burgeoning growth of Internet use in China helped drive advancement in technologies that now allow Chinese characters to be supported in a variety of software, including those used for blogging. So from the start, Chinese users have participated in developing the blogosphere.

In fact, scholar and expert on Chinese Internet use, Rebecca MacKinnon had this to say after attending the first CNBloggerCon, held in Shanghai during November 2005:

> Web2.0 is potentially a very Chinese thing. One of the most important words in the Chinese language is "guanxi." It means "relationship." Whatever you think about the term "Web2.0," the point is that social networking and relationship-building are at the core of today's most exciting web innovations. . . . You are nothing in China—and can accomplish very little—without a good "guanxi" network. Expect Chinese internet users to seize upon Web 2.0 tools as a way to expand and deepen their human relationships, enhancing both personal lives and businesses. Expect Chinese users to build new tools that suit their own preferred ways of communication. The Chinese are likely to have a growing impact on the evolution of web applications. (MacKinnon 2005)

On the Internet, because performing identity depends far more on self-description than on presentation of direct "evidence," every blogger must make choices about what to reveal and make further decisions about how to illustrate those choices. Looking at these choices we can explore what criteria seem important in defining identity and in what circumstances and in what ways bloggers and their readers choose to exhibit their belonging to a particular culture or ethnicity. Though the Internet was initially hypothesized as a realm naturally free from sexism, racism, and other forms of discrimination (Turkle, 1995), it now seems that users are quite concerned with identifying themselves and others in terms of ethnicity, gender, interests, and so on (Nakamura, 2002). Does this then mean that we are bound to the stereotypes of race and ethnicity that plague every society in one form or another? Or instead are we headed for a homogenized blogosphere in which everyone writes to a Western standard? Perhaps neither.

Recent research on the characteristics of Internet users and of the effects of Internet use has been suggestive: those who socialize the most online tend to be more sociable in general, and Internet users have been shown to be slightly more optimistic and slightly more open to diversity than nonusers (Boase et al., , January 25, 2006). These characteristics could make online communities more likely sites for positive experiences of intercultural communication. In addition, while identity categories are being reenacted online, meaningful participation in discussions often depends on knowledge of a culture/country, awareness of national concerns, and a general feeling of goodwill toward the culture/country around which the community is focused.

As blog content and behavior both reflect and challenge the status quo, certainly the blogosphere is developing in ways that are far more complex than was initially expected (Lyons 2005). As more individuals enter the global community via blogs, studying these developments is essential to our understanding of identity, as well as the impact of this kind of digital communication. The blogosphere is a challenging area to study because it is so vast and changes so rapidly. New bloggers join every day; seemingly established blogs and communities come and go. Establishing patterns of influence will take many more years of broad observations, but looking at even a few long-running blogs and communities can be informative. This chapter traces the development of five individual blogs and explores how issues of nationality are addressed in the blogs, and what this might reveal about blogs as a carrier of Western cultural norms or as site for performance of Chinese identities.

Sinosplice

Sinosplice began in April 2002 when John Pasden, a grad student in linguistics who had gone to China to teach English and learn Mandarin, decided he needed a better way to stay in touch with family and friends than the mass emails he had been sending out. He soon realized that his blog was part of a small but growing number of blogs kept by expats in China, and he began supporting the growth of that community by sharing his server space. John is not Chinese, but his blog was one of the first noncommercial sites to encourage blogging in China, and he maintains one of the most comprehensive lists of Chinese blogs (http://www.china bloglist.org/). He also initially hosted some of the earliest and most widely read Chinese bloggers, including two discussed here, Leylop and Andrea Leung of T-Salon, and still hosts a number of Chinese blogs at his site. For these reasons, knowing something about his blog will help us understand the larger Chinese blogosphere.

John's blog has always been mainly a chronicle of his experiences in China and aimed to be a resource for others who might be interested in teaching English or traveling in China. He frequently posts entries that he feels will combat stereotypes often held by Westerners about China, and these may lead to a lively response from readers. For example, in March 2004, he posted an excerpt from a Chinese blog that he translated, about a Chinese woman's difficult experience telling her parents that she was dating a white American. John then shared his own very different experience meeting the parents of his girlfriend, who is also Chinese. He ends with the following points:

> All I'm trying to say here is:
> 1. China is such an incredibly varied place; you get all kinds of people with all kinds of life circumstances and outlooks.
> 2. Shanghai is a singular phenomenon in China. There is no city like it, for so many reasons.
> 3. I am really incredibly lucky. (March 4, 2004)

John's post generated 96 comments, the most recent added on May 19, 2007. Within these posts we see a range of opinions about whether or not various stereotypes (identified by the writers) are true:

> Chinese women only marry white men to get a visa (sic)
> White men who pursue Chinese women have a "Suzie Wong" complex
> Chinese parents generally hate the idea of a white son-in-law

Though the writers agreed about the existence of the above three stereotypes and argued over their accuracy, one stereotype was notably absent from the thread: that Chinese people are too foreign, and non-Chinese can never really understand them. Although individual posts sometimes were inflammatory, most writers engaged in a serious discussion and shared their own experiences. In all of the posts, writers identified their own ethnic origins and the source of their experiences, if they were sharing them. Most participants in the debate seemed well aware that people's opinions were affected by both their own experiences and culture; they basically agreed that there were sincere and insincere people in every culture, and no one rule would always hold. Over the three years the discussion continued, we also learned that the young woman whose blog was quoted did eventually marry her boyfriend, who ended up getting along very well with her parents and relatives, once they'd met in person.

Though his posts have generally been descriptive and informative about China and are often fairly opinionated, they are also apolitical and give little hint of John's views of the Chinese government or its policies. This has been deliberate choice on

John's part, and even led eventually to the departure of some blogs he had hosted, including T-Salon, that he felt were too political. However, John has developed and hosts the "adopt-a-blog" program to help Chinese bloggers evade government censorship, so we must assume he believes that bloggers who do want to write more explicitly and critically about politics should be able to do so. Though I am looking primarily at the way bloggers in China perform their own ethnicities or nationalities, these cannot help but sometimes cross into more generally political topics and when it comes to blogs, whether or not people in China write about politics has been tied to a Chinese identity by some writers, including another prominent blogger, Isaac Mao.

Isaac Mao

Isaac Mao has been blogging since early 2002 and shortly after starting a blog he created one of the first Chinese blog sites, CNBlog.org. He is also extremely active in the open source movement and was part of the team that developed "Creative Commons China," part of the Creative Commons organization founded by Stanford Law Professor Lawrence Lessig. Mao typifies most of the early bloggers in China; before starting his own company, he worked for Intel China and is both an experienced programmer and an experienced netizen. He started his blog site in order to encourage grassroots publishing in China and to share his thoughts more generally. In other words, to help open the door to Chinese writers and others who wanted to publish their ideas without dealing with the bureaucracy of publishing corporations or government censorship. Since opening his blog site, Mao has commented extensively on the role of blogs in China with regard to Chinese identity, censorship, and other issues, and is now regarded as one of the leading authorities on these matters still working in China.

As a leading blogger and a member of the Internet industry, Mao has also organized the CNBloggerCons that have brought together Chinese bloggers to discuss many issues relevant to them all. These conventions have gotten considerable attention in the media and many of his speeches are quoted or paraphrased in Western media. For example, at the second CNBloggerCon, held in late 2006 in Hangzhou, Rebecca MacKinnon reported that Mao claimed that blogs were crucial in addressing Western stereotypes about China:

> Mao talked about how the international media covers China in a limited way that many Chinese people feel is biased and unfair. Blogs, he believes, are an important way for people to take matters into their own hands and represent themselves. He eloquently explained why it's important for Chinese bloggers to think more globally, and consider their power as media creators to make a difference in the way that the rest of the world understands them. (MacKinnon, cnbloggercon/index.html)

Mao has thought a great deal about how blogs, wikis, and other software affect the way people think and interact; these topics form the bulk of his posts in his English blog. While he does not seem to ever worry that blogs are somehow squeezing non-Western writers into a Western cultural framework, he talks at length about how Chinese bloggers write from a different experience than their Western counterparts, and experience that hinders their thinking (Mao, "Free Thinking," 2007). He feels that blogs will help Chinese users overcome a habit of distrust learned through years of daily insecurity.

> Due to political reasons from the past, Chinese people don't like to share their viewpoints, they tend to hide themselves so they don't trust each other. But blogs help people to trust. And unlike the traditional media, this is not controlled by anyone. (*ManilaTimes*)

In fact, in 2005, a survey carried out by Tsinghua University professor Jin Jianbin and Sohu.com found that 77 percent of bloggers in China say their intent in writing is to express their feelings (Young, 2005) . It seems people in China are hungry for such an outlet. The next two bloggers, Wang Jianshuo and Leylop, discussed in the following sections, use their blogs for personal expression but along the way encountered the same problem that Mao mentions above—that a biased image of China is held abroad, and that their blogs can help to counter that image.

Wang Jianshuo

Another early adopter, Wang Jianshuo began his blog in early 2002, writing mainly about his life in Shanghai. At that time, he worked (and still does) for Microsoft, and was engaged—he is now married. Wang has always kept his blog fairly narrowly focused on life in Shanghai, his travels, day-to-day events in his life, and his personal thoughts on various topics but especially on blogging itself. In fact, his earliest entries are quite technical, more notes to himself or friends than any kind of outreach. At the same time, Wang was thinking of a larger audience. In a July 2004 entry, he explains his motives:

> Back in Jun of 2002, my friend Wu Hao told me a story about the monk and the well. It was about two monks living in the same mountain. Both of the two monks go to river far away to get water. One year later, one monk did [not] have to go to the river while the other had to go. The reason is, the smarter monk used the year to dig a well by himself everyday, while the other one did nothing. It reveals the importance of accumulation.
>
> I thought to write something every day may be a good idea. Then I heard the concept of blog and I started. Maybe one day, even I leave Microsoft, the company brings me

much honor and opportunity, I am no longer an employee of a famous company, I am still the writer or a famous blog. I won't start from scratch. As the simplest example, now, if I post my resume on my blog—a site with many readers, I may have more chances than others. I learnt from private emails that some readers are executives of big companies. (20040701_i_started_the_blog_to_post_resume.htm)

Wang Jianshuo has continued to blog at least every few days, and he continues to focus on mostly personal and sometimes helpful-to-others topics. Though his blog is written in so mild a tone that it might even be called bland, it gets many visits from other countries. Wang himself attributes this in part to his providing practical information about visiting Shanghai and China, and in part to his inclusion of personal details. This view is evident in the comments to every entry, which are largely positive and full of thanks.[3] However, even Wang gets flamed sometimes as he describes in a September 2004 entry:

Why a Chinese uses English to Blog?

It is common comment. Someone said something really bad about this blog, because it is in English. Many people take France as a model of anti-English countries. I have to accept this. It is bitter sometimes. Well. Let me get back to the question, why I choosed to use English in this blog?

It was just for fun at the very beginning. When I started, who knows what the site would be look like. Just as the URL of this site, I used http://home.wangjianshuo.com for simple reason—the computer was at my home. After that, I found the gap of communication and the hopeless visitors in Shanghai or planning to visit Shanghai. So I continued. It does not mean I don't love my country or to show off—as some commented. (20040915_negative_comments_for_this_blog.htm)

Notice that Wang is condemned for blogging in English, not for blogging at all. After the above post, many readers sent supportive comments validating his choice to blog in English and thanking him for opening a window on Chinese life. His readers are also rather protective of Wang; they were quick to point out how slanted an article that was printed six months later in *Wired Magazine* was; the magazine that profiled Wang presented him as an example of the new China (Margonelli, 2005). Wang agreed but was extremely gracious about assuring readers that *Wired* had checked their facts; they had just misinterpreted his feelings. In fact, the *Wired* article reveals that a great deal of stereotyping may occur in the mind of a reader (in this case the *Wired* staffer responsible) rather than in the actual blog itself.

And this may be a crucial point: typical blog readers construct images of the blogger that appear to be based far more on their own preconceptions than on what the blogger has actually said. Wang himself has recognized the extent to which stereotypes still constrain the understanding even of people who fervently want to learn about China. After returning from a visit to Silicon Valley, he commented about this at length in a June 2006 post, saying that "People in U.S. Don't Understand China" and that "People in China Needs to Understand American as Well (June 13, 2006)." As he goes on to say:

> On the facts part about China, what people in U.S. get is typically correct. The problem happens in the reasoning logic.

> For example, people understand the terms of law in China, but are lack of information (sic) about how laws are enforced in China. So they interpreted in U.S. way.

Concluding that many bridges are needed between our two countries, Wang set out to more explicitly become one, creating a subcategory in his blog devoted to East-West discussion, and going back through his blog to tag the relevant entries. Though many of the bloggers discussed so far mention how their blogs have changed their lives by allowing them to meet many new people from all over the world, few experienced so material a change as Leylop.

Leylop

Leylop started blogging in 2002 while she was majoring in history at Zhejiang University. Leylop was inspired by Sinosplice and decided to try her own English blog as a way to practice writing in English. At first Leylop wrote entries that were not so different from what we still see when college students start blogging in the United States; she was bored with school; she was listening to this or that; she had seen something interesting around town. Sometimes she would post pictures as well, such as street scenes, flowers, old men sitting and playing weigi. She began to attract readers, often people who had visited China, or were planning to, or teachers who found her blog through Sinosplice. Nearly all of these readers responded positively, commending her English language skills and responding as well to her commentary. Readers also praised her for being open to new experiences merely because she tried blogging, which suggests they had expected her not to be open or tech-savvy; trying a blog doesn't in itself generally earn praise for Western bloggers. Fairly quickly though, Leylop began adding more pictures and sharing more decided opinions about China, music, school, local culture, and so on.

> Obviously, the Chinese language teaching in China is a failure. Take Hangzhou for instance, just several (less than 5, maybe) Chinese schools are here and most of them require enrolling for full-time learning. Why don't we open more Chinese schools or training classes as we did for learning English? Why don't we create more chances of being understood by letting others speak our language? (December 29, 2002)

Leylop began to find an even wider audience and was linked to many other blogs. She also began to encounter more resistant readers. Of course, some of her readers were trolls as any blogger may encounter from time to time. Leylop sometimes engaged them for a while but mostly ignored them or deleted them. Some people objected to her comments on China; interestingly some of these respondents were Chinese, but some were American and their objections did not divide neatly along national borders. The Chinese might object to her criticizing China, but they often agreed with her views. They might question her choosing to write in English (though she soon was blogging in Chinese as well) but most did not seem to care about this. The American critics expressed even fewer consistent opinions. Some simply wanted to complain about China and used her blog as a platform. Some thought she was being prejudiced against the Chinese herself. Some thought she had been co-opted by the West because not only was she writing in English, but she was downloading and listening to both Western music and Western films. They mourned the erosion of Chinese culture. Not surprisingly, Leylop found this last view rather insulting, but she didn't assign these disagreements to cultural differences (January 2003). In a 2006 interview posted on the China Radio International Web site, Leylop makes the following comment on cultural clashes:

> I do get emails from Westerners voicing different opinions. But Chinese readers send me emails with opinions different from my own, too. So I don't think the difference in our thoughts is due to cultural differences; we are, after all, different people. (CRIEnglish.com)

As Leylop continued talking with a now global audience, she found a new reason to be interested in school and to develop some goals for herself; she decided she wanted to travel. It seems that one effect of blogging had been to widen Leylop's horizons, at first beyond her home city, then beyond China. But beyond creating the desire for travel, blogging aided her in achieving this goal as well. Through her blog, Leylop met people who helped her with both employment to raise money and arrange her travels, with some even hosting her during visits or becoming a travel partner, on occasion. Leylop documented her travels with entries posted along the way and added many photos as well. Her photography skills improved even more rapidly, and this, along with her travelogues, attracted even more visitors to the blog.

Not surprisingly, as Leylop journeyed, her comments grew more nuanced and more critical. Not in the sense of being negative but more questioning and philosophical; she began to think more deeply about the places she saw and people she met. For example, after traveling in Daramsala (May 2005), she expressed quite a complicated view of Tibetans and the relationship between Tibet and China. Shortly thereafter she commented in July 2005 about how her view of foreigners had changed: "Actually there's no foreigner in my eyes anymore, we're just people from different places" (July 10, 2005). At the same time, her view of her own country had also changed. Leylop doesn't usually comment much on politics, but she posted two entries that make clear how travel and exposure to so many other cultures and views have affected her. The first of these comments was on the anti-Japanese protests that occurred in China during the spring of 2005. Leylop wrote:

> Yeah, it's a shame that Japanese government doesn't have the courage to face the history, but 60 [years] later some Chinese still hate Japan and Japanese because of the war, well, that's way too much. And let's take a look at Tibet. Do Chinese history books ever mention the invasion of Tibet and Free Tibet movement? Do Chinese government ever talk about what happened in Tibet fifty years ago? Never ever. Chinese history books are not that objective on many issues themselves, what's point of blaming others? (April 13, 2005)

After receiving quite a few negative comments from Chinese readers, and some that were truly vicious, Leylop finally said:

> Almost one year of travelling outside China makes me think a lot. I don't want to be too cynical to my country, but I have to reconsider China as a united country and question Han-Chinese's policies on other ethic groups in different perspective. After I wrote "No anti-Japanese," I've got several hate mails from some angry Chinese. I don't want to argue with these people, we're just different. (May 29, 2005)

Here we see the same conclusion reached by the other bloggers: that most clashes are not caused by cultural differences but merely by the differences that might arise between any two people, even if they share cultural and ethnic identities. At the same time, shifting the focus away from cultural stereotypes allows both bloggers and readers to see they don't all hold the same values, or even if they do, they may express them differently, that understanding many opinions often does depend on understanding the context in which they have been formed and expressed. Finally, Leylop says quite clearly that her blog is just about her and who she is:

> Someone left a harsh comment on my last entry in Chinese. Usually I don't justify what I write, yes, sometimes I am inconsiderate, greedy and mean, but that's who I am and that's it. This is not a place to represent what a model citizen I am or how important Chinese values are. (Leylop.com, June 1, 2006)

Here she reiterates what she said in one of her earliest posts that she was impatient, abrupt, and not nice, among other things (December 31, 2002). Though her readers might disagree, we can't help but note that if nothing else, she is consistent, and that all of the characteristics that might be criticized as not being Chinese, or as being too Western, have in fact been part of her personality from the beginning. So far all of the bloggers seem to agree that these online journals are primarily places to express themselves individually, and their writing and other evidence suggest that though this view might seem inherently Western, it is not, and has by no means been imposed. We see from the comments of the bloggers themselves that they do not perceive their personal focus or a particular way of expressing themselves as being caused by the blog and don't seem to see themselves as acting or writing in an "un-Chinese" way either. And of course this is not surprising since as noted earlier, Western notions of authorial purpose and of good writing arrived in China quite a bit earlier than blogs did and are by now integrated into many schools, especially in the classes teaching English. This is not to say that cultural transfer has not occurred, but only that blogs have not been responsible for this particular instance, and, more positively, that Chinese bloggers are not naive about the different cultural values that inhere in different ways of writing and of approaching the Web.

But the last blog to look at is a departure in that it contains almost no personal details and is not used primarily as a diary. On the other hand, it does aim to break down barriers in understanding.

Andrea Leung and T-Salon

Andrea Leung is a native of Hong Kong who began writing a blog from Hangzhou called T-Salon in early 2002. Like many of the early Chinese bloggers, she had a fair amount of technical expertise already and was comfortable with computers. Unlike other bloggers, she was not as interested in keeping a diary; she instead wanted to use the blog as a repository of useful or interesting things she found online that she could save with some comments and tags. Her academic interest has always been in media, and she began by first by simply collecting books she was reading, newspaper articles she found, or speakers she saw, generally all connected to media in China. Occasionally and early on when the blog was still new, she posted about daily life, but she found that her family and friends weren't following as she had hoped and that she had to email them all anyway.

Now Leung's blog has become mainly an outlet of information for anyone interested in developments in the world of Chinese blogs, and she doesn't offer as much of her own commentary as the others do, and generally seems to get fewer comments. She is currently based in Vancouver and is still posting to T-Salon, as well as writing for Global Voices and is involved with some of Isaac Mao's projects. While in all the above ways she is very active and voices her opinion, she spends much less time posting about her own Chinese identity or Chinese national identity as such. This is in fact a reason why her blog, more than the others, may relieve fears of cultural imperialism. However it also complicates the rosy picture some of the other bloggers seem to hold of blogs as instruments to create greater understanding, because Leung simply doesn't engage very much with readers or seem interested in connecting with them. Of all the bloggers discussed here, she is the only one who has not explicitly expressed any desire to create greater understanding and not much desire to express her own views or experiences.

In not feeling any need to address issues around her Chinese identity, Leung demonstrates the extent to which she is comfortable with blogs; she takes them for granted and uses them as tools. She also doesn't seem to be deliberately trying to act as a bridge in the same way as the others—by making personal connections. Instead, Leung acts to make information more accessible to her readers, both providing articles from the Western media for Chinese readers and translating Chinese articles for Western readers. Thus her blog is a valuable resource for readers but does not create the same feeling of community as other blogs discussed here. In this way Leung also undermines stereotypes about Chinese collectivism; nothing in her blog expresses a collective feeling any more than a Westerner's blog. In fact, none of these blogs lines up consistently with the characteristics attributed to Chinese people or Westerners.

Bridges or Breaches?

Though we can see commonalities between the bloggers here in various categories—motive for writing, background with technology, choice of general topics, conclusions drawn about people, and so forth—none of these commonalities lines up neatly with national origin. This lack of a clear pattern accords with the views of authorities such as Mao and MacKinnon in suggesting that blogs have not served as a major conduit for Western cultural imperialism. They clearly are affecting Chinese culture, just as they are affecting the cultures of any country in which they are popular. In some cases they do seem to lead to greater understanding and a more cosmopolitan view of people living in other countries and a sense of identity that is less dependent on reference to a national culture. Blogs however can also be places where flamewars erupt, or simmer. Though this rarely occurred in the blogs profiled here, other research suggests that in Chinese language blogs, attitudes toward international readers are far less tolerant. So the optimism evinced by most of these five bloggers may be conditioned by their attracting audiences that are more likely to be open-minded and to react favorably to any attempt at intercultural communication. Certainly further study would help to clarify the extent to which blogs change or reflect cultural values.

Returning to Rebecca MacKinnon's notes about the CNBloggerCon 2005:

> Another thing about this story: it's not so much about what the internet is "bringing" to the Chinese, or how the internet is coming in as an outside force and "changing

China." The real story is about how Chinese users are taking the connectivity, tools and applications, internalizing them, and making them their own.

This is what we really need to explore, if we wish to understand how blogging works. We need to ask not whether Chinese people blog like Americans do, or whether they don't; rather we need to ask how people blog, wherever they are, period (Su et al., 2005). Studying this will help us learn about how different circumstances affect the way people use blogs.[4] Certainly years of living under the totalitarian regime of the Central Government have led to certain behaviors that we might identify as typically Chinese, as Isaac Mao noted when he spoke at the CNBloggerCon: "It's only natural for human beings to express themselves and share their views. But the Chinese people have been repressed for so long and they have always kept silent" (ManilaTimes.net). However we know that the U.S. government has surveilled email they think may shed light on terrorist activity, and during May 2007, the company that operates LiveJournal purged thousands of journals based only on interest keywords—a tactic quite familiar in China.

Further, Chinese bloggers are not forced to follow Western modes; there are many Chinese BSPs and some people involved with these, such as Isaac Mao, have launched other Web services that are not only popular in China but are now sweeping across the Net, demonstrating that culture and technology flow both ways.[5] The trouble with talking about bridges or breaches is that both images are based on China being separated from the rest of the world; whether by a moat or a wall, it makes no difference because we are still using terms that represent China as completely foreign and inviolate, conveniently ignoring millennia of interaction with other cultures. As the blogosphere expands in China and we continue our exploration, we would do better to think of it as just another neighborhood, not so different from our own.

Notes

1 A more general claim for this possibility was advanced as early as 1998 in Kim, 1998.

2 This fear is not unfounded since this dynamic has certainly been seen in other media such as film, music, and software, as explored by Virilio in earlier work. Further, the rise of English as a global language has also led to fears of cultural imperialism that are not without basis; local dialects are disappearing all over the world and this is especially evident in China where Putonghua and English are often both mandatory from kindergarten onwards (Virilio, 1995; Crystal, 2003; Turner, & Acker, 2002

3 Wang Jianshuo was tracking his comments until summer 2006 at least, and through that page (Wang, 2006) it is easy to see that he has many regular commenters whose posts are always constructive. Negative posts (as opposed to those that just disagree) are rare enough that they usually lead to blog entries from Wang and can be easily searched.

4   Some recent papers, Tsui, 2001 and Honan, 2004 have focused on how government monitoring and suppression have affected blogging practice in China, but my observations suggest that Su et al. (2005) are correct in concluding that in fact bloggers around the world work in very similar ways. In fact, Tsui (2005) also takes a broader view in his later research.

5   One good example is Anothr, an aggregator website with whom you may register, choose blogs and news sites you wish to follow, and have their headlines delivered via a chat window—saving you from having to visit each site or cluttering your own page or inbox with feeds.

References

Bausch, P., Haughey, M., and Hourihan, M. (2002). *We Blog: Publishing Online with Weblogs*. New York: John Wiley & Sons, Inc.

Blood, R. (2000). *Weblogs: A History and Perspective*. Retrieved May 2007, from *Rebecca's Pocket* Website: http://www.rebeccablood.net/essays/weblog_history.html Sept. 9, 2000.

Blood R. (2002). The Weblog Handbook. New York: Basic Books.

Boase, J., Horrigan, J., Wellman, B., & Rainie, L. (2006, January 25). The Strength of Internet Ties: The internet and email aid users in maintaining their social networks and provide pathways to help when people face big decisions. *Pew Internet & American Life Project*. January 25, 2006. Retrieved May 2007, from http://www.pewinternet.org/PPF/r/172/report_display.asp

CRIEnglish (2006, April 1). *Backpacker and blogger—Leylop*. Retrieved May 2007, from http://english.cri.cn/974/2006/04/01/63@70092.htm

Crystal, D. (2003). *English as a global language*. New York: Cambridge University Press.

Honan, M. (2004, June 4). "Little red blogs". *Salon*. Retrieved May 2007 from http://dir.salon.com/story/tech/feature/2004/06/04/china_blogs/

Jianshuo, Wang (2004) "I STARTED THE BLOG TO POST RESUME" retrieved Sept 2006, from http://home.wangjianshuo.com/archives/20040701_i_started_the_blog_to_post_resume.htm

Jensen, M. (2003, September/October). "A History of Weblogs." *Columbia Journalism Review*, 42(3), 22. Retrieved May 2007, from *http://cjrarchives.org/issues/2003/5/blog-jensen.asp*

Jin, J. (2007). The stickiness determinants of blogging: from the perspective of user acceptance of information technology. Paper presented at tthe 57th Annual Conference of the International Communication Association: Creating Communication: Content, Control and Critique, San Francisco, USA, May 24–28.

Kim, S. (1998, Fall). Cultural imperialism on the internet. *The Edge: The E-Journal of Intercultural Relations, 1*(4). Retrieved May 2007, from http://cms.interculturalu.com/theedge/v1i4Fall1998/f98kim.

Lee, R., & Wong, S.-L. C. (Eds.). (2003). *AsianAmerica.net*. New York: Routledge.

Lenhart, A., & Fox, S. (2006, July 19). Bloggers: A portrait of the internet's new storytellers. *Pew Internet & American Life Project*. Retrieved May 2007, from http://www.pewinternet.org/PPF/r/186/report_display.asp

Leylop (June 1, 2006). *Leylop.com*. Retrieved May 2007, from http://www.leylop.com

Leung, A. *T-Salon.net*. Retrieved May 2007, from http://www.t-salon.net

Li, Xiao-ming. (1995). *"Good Writing" in Cross-Cultural Context*. Albany: SUNY UP.

Lyons, J. K. (2005, Fall). Media globalization and its effect upon international communities: Seeking a communication theory perspective. *Global Media Journal*, 4(7). Retrieved May 2007, from http://lass.calumet.purdue.edu/cca/gmj/fa05/gmj-fa05-lyons.htm

MacKinnon, R (2005). "Chinese Bloggers: 'Everybody is Somebody.'" Retrieved January 2007, from http://rconversation.blogs.com/rconversation/2005/11/chinese_blogger_1.html

MacKinnon, R (2006). "Blogging between China and the rest of the world" Retrieved January 2007, from http://rconversation.blogs.com/rconversation/2006/10/blogging_betwee.html

MacKinnon, R. (2007). *RConversations*. Retrieved May 2007, from http://rconversation.blogs.com

ManilaTimes. Blogger leads China to free-thinking revolution. Retrieved May 2007, from http://www.manilatimes.net/national/2007/apr/03/yehey/techtimes/20070403tech1.html

Mao, I. (2007). *Free thinking weaves new social media (in China). Chinese Grassroots Media: In Search of Free Thinking*. Seminar held by the Journalism and Media Studies Centre, Hong Kong University, Hong Kong. Retrieved May 2007, http://rconversation.blogs.com/rconversation/2007/04/isaac_mao_and_m.html

Mao, I. *Meta*. (2004). November 2004 Retrieved May 2007, from IsaacMao Web site: http://www.isaacmao.com/meta/

Mao, I. (2005). Keynote given at CNBloggerCon, November 5–6, 2005. Translated, transcribed and posted on the *Chinese Blogger Conference* blog by Angelo Embuldeniya. Retrieved May 2007, from http://cbc2005.wordpress.com/2005/11/05/first-session-by-issac-maos-part-2/

Margonelli, L. (2005, April). China's next cultural revolution. *Wired* (13.04). Retrieved May 2007, from http://www.wired.com/wired/archive/13.04/china.html

McChesney, R. . (2001). "Global Media, Neoliberalism, and Imperialism." *Monthly Review*, 52.10.

Mortensen, T., & Walker, J. (2002). Blogging thoughts: Personal publication as an online research tool. *Researching ICTs in Context*. Online conference Archive of Intermedia , Oslo, Norway. Retrieved May 2007, from http://www.intermedia.uio.no/konferanser/skikt-02/docs/Researching_ICTs_in_context-Ch11-Mortensen-Walker.pdf

Nakamura, L. (2002). *Cybertypes: Race, ethnicity, and identity on the internet*. New York: Routledge.

Pasden, J. *Sinosplice*. Retrieved May 2007, from http://www.sinospice.com

Pew Internet & American Life Project (2006). Bloggers: A portrait of the internets new storytellers. Retrieved December 2006, from http://www.pewinternet.org/pdfs/PIP%20Bloggers%20Report%20July%2019%202006.pdf

Qiang, X. (2004, November 24). The "blog" revolution sweeps across China. Retrieved May 2007, from *New Scientist* Website: http://www.newscientist.com/article.ns?id=dn6707

Robinson, J. P., Neustadtl, A., & Kestnbaum, M. (2002, Summer). The online "diversity divide": Public opinion differences among internet users and nonusers. *IT & Society*, 1(1), 284–302. Retrieved May 2007, from http://www.stanford.edu/group/siqss/itandsociety/abstract.html

Su, N. M., Wang, Y., Mark, G., Aiyelokun, T., Nakano, T. (2005). *Proceedings of the Second International Conference on Communities and Technologies (C&T 2005)*. The Second Communities and Technologies Conference, Milano, 2005, Springer, The Netherlands.

Tsui, L. (2001). *Big mama is watching you: Internet control and the Chinese government*. Unpublished M.A. thesis, University of Leiden, The Netherlands. Retrieved May 2007, from http://www.lokman.nu/thesis

Tsui, L. (2005). The sociopolitical internet in China. *China Information: A Journal on Contemporary China Studies, 19*(2), 181–188. Retrieved May 2007, from http://www.lokman.org/sociopoliticalinternetinchina.pdf

Turkle, S. (1995). *Life on the screen: Identity in the age of the internet.* New York: Simon and Schuster.

Turner Y. & Acker A. (2002). Education in the New China. Surrey, UK: Ashgate.

Virilio, P. (1995). "Speed of Information: Cyberspace Alarm." *CTHEORY, a030.* Retrieved May 2007, from *http://www.ctheory.net/articles.aspx?id=72*

Wang, J. (2006). *WangJianshuo.com.* Retrieved May 2007, from http://home.wangjianshuo.com/

Wong, J. I. (2005, November 22). In blogs we trust. *South China Morning Post.* Re-posted at AsiaMedia, hosted by UCLA. Retrieved May 2007, from http://www.asiamedia.ucla.edu/article.asp?parentid=34176

Young, Liu (2005) Chinese Blog Survey Reveals Online Habits—ChinaTechNews. Retrieved Jan 2006 from http://chinadigitaltimes.net/2005/10/chinese-blog-survey-reveals-online-habits-chinatechnews/

Zuckerman, E. (2005, September 16–17). *Meet the bridgebloggers: Who's speaking and who's listening in the international blogosphere.* Paper presented at The Power and Political Science of Blogs, University of Chicago, Sept16–17, 2005. Retrieved May 2007, from http://ethanzuckerman.com/

CHAPTER FIVE

Blogging IN Russia Is NOT Russian Blogging

KARINA ALEXANYAN AND OLESSIA KOLTSOVA

What Is "Runet"?

The "Russian Internet" or "RuNet" is an ambiguous label. It can be understood nationally and geopolitically as the "Internet in Russia," or culturally and linguistically as all Russian Web resources worldwide. These are two very distinct, albeit overlapping, entities.

The Internet in Russia is limited to Web resources within the national domains .ru (and .su) servers territorially based in the Russian Federation and data that originate on these Russian networks. The Russian language Internet, on the other hand, includes Web resources created by an enormous global diaspora of émigrés and residents of former Soviet republics. If one includes Russian as a second language, over half the world's approximately 255 million Russian speakers live outside of Russia (Ethnologue, 2007).

This chapter explores the Russian language Internet and Russian language blogging, with a particular focus on how issues of culture and identity manifest themselves online—in the emergence of transnational networks, as well as the assertion of traditional geopolitical concerns.

We focus on the American blogging and social networking site LiveJournal (LJ) (www.livejournal.com)—which is also one of the largest aggregates of Russian language activity online. With over half a million Russian users, LiveJournal is a key online resource for the Russian community—both within Russia and abroad. The history of LiveJournal is so intertwined with the evolution of the Internet in Russia that its initials (*Zhe Zhe*) are synonymous with blogging.

We look at how LiveJournal is used by Russians both within and outside of Russia, including the collusion of elements (technical and cultural) that have led to the site's Russian appeal, the shifting elements that constitute a "Russian" user online, and the characteristics that differentiate Russian LiveJournal users from their American counterparts. We explore how the growing Russian LiveJournal community, both national and transnational, is used by global commercial interests, with a particular focus on the controversy and discussion surrounding the recent licensing of LiveJournal's Russian segment by a Russian company. In this context, we examine the value conflicts that can arise when traditional territorial understandings are imposed upon transnational communities, illustrating the persistence of geographic values on the Internet and the continued significance of "place" in the virtual world. The chapter combines contemporary scholarship and journalism on Russia and the Internet with an analysis of online posts and other primary written sources, and a pilot series of authors' interviews with American and Russian LJ users.

Internet and Society

The various popular approaches to the Internet and its influence within Russia may be arranged into two major themes—the first subsumes the Internet into larger discussions of what constitutes "Russian culture" and the second offers conflicting views as to the Internet's role vis-à-vis politics and the state.

Internet and Culture—Slavophiles versus Westerners

The first theme roughly corresponds to the broader and much older discussion of the "Russian way" between so-called Zapadniks and Slavophiles. The Slavophile approach highlights Russian cultural uniqueness and the continuity of folk traditions, whereas Zapadniks emphasize technology and Westernization/ Internationalization (Schmidt & Teubener, 2006:17). The Zapadniks accept an external gauge of progress, and their philosophy focuses on ways in which Russia can "catch up," primarily in a material and socioeconomic sense. The Slavophile perspective, on the other hand, is culturally isolationist and creates its own gauge, in which Russia lies in social superiority of the West. In this narrative, "Russia does not position itself as a 'peripheral receiver' of Western cultural messages but rather as . . . an alternative source of cultural production, whose content is primarily spiritual rather than material" (Pilkington et al., 2002:13).

In terms of the Internet, the Slavophile approach assumes "a typical Russian mentality" and projects it on the new medium, highlighting those characteristics that resonate with stereotypes of the Russian national and ethnic character. In this context, "the Internet is no longer a Western import but is seen as something genuinely Russian," and the characteristics and parameters of "RuNet" are determined by cultural identity, above and beyond shared language (Schmidt & Teubener, 2006:17). So, for example, while in the United States, the emerging Internet was illustrated by uniquely American metaphors, referencing a new "Wild West" or "Frontier," as well as an "information superhighway," in Russia, it was conceptualized in a more grassroots fashion, as place for dialogue akin to the Soviet-era domestic kitchens, where traditionally so many philosophical and political debates took place.

In another example, Slavophile and Russian religious philosophies emphasize those features of the Internet that resonate with their core values of "collaborative ethics and aesthetics, spiritual unity and collectivism, creativity and freedom" (Schmidt & Teubener, 2006:18). As a result, those characteristics of the Internet that promote collectivism, de-hierarchization, and egalitarianism—such as interconnectivity, interactivity, and social networking—are interpreted as innately Russian, resonating with Russian cultural and mental patterns. Hence the introduction, in the late 1990s, of the term "Electronic Sobornost," alluding to a "spiritual unity, which nevertheless allows for individual creativity and self-realization" (Schmidt & Teubener, 2006:18). The term "Sobornost" is based on the Russian word "sobor," which, in modern language, primarily means "cathedral," but has its roots in archaic references to gathering and meeting. The Slavophile interpretations of the Internet illustrate the process of "glocalization"—adjusting a global or foreign phenomenon to local/national traditions and interests, including the extreme view of strong nationalists and radical isolationists.

As opposed to this "Slavophile" perspective, most Russian businesses and commercial interests favor a "Westernization/Internationalization" approach, with the goal of integrating into global networks and adapting and assimilating Western standards. This perspective is more quantitative, and the characteristics of RuNet are understood based on external factors such as implementation rates, infrastructure, financing, and media control/regulation (Schmidt & Teubener, 2006:17). This outward-looking perspective is stimulated by the current sociopolitical climate. "Against the background of ideological 'overload,' of growing nationalism within the country, the idea of a specific development of the Russian Internet is rejected. . . . [This perspective] denies the notion of national specifics in global technologies. On the contrary, the Internet offers the possibility to escape narrow, national contexts . . ." (Schmidt & Teubener, 2006:19). As one of Schmidt and Teubener's sources, a graduate student, asserts, the Internet is "a way into the outside world, rather than a sort of connection with others in Russia" (Schmidt & Teubener, 2006:19).

What unites the two groups, apart from a mix of descriptive and normative/tele-ological elements, is "the reference to Western influence which should be either absorbed and assimilated, or resisted, but cannot be ignored" (Koltsova, 2007:55).

Internet and Politics—Pessimists and Optimists

The second theme focuses on the relationship of the Internet to the state and, despite features unique to Russian politics, does not differ much from Western hopes and fears regarding Internet's impact on democracy, civil society, and communicative freedom. Both "Optimists" and "Pessimists" admit the huge influence of the Internet on society but argue on its quality—on the extent to which the Internet can remain independent of political/state influences, and on the role of business/commercial interests. Optimists, such as Russian media researcher and Internet entrepreneur Ivan Zassoursky, point to the Internet's decentered and unregulated character and posit it as an effective counterforce to the existing political regimes and the mainstream media they control. In this perspective, the Internet represents a bastion of freedom and a link to the outside world, especially in the face of rising media consolidation. "At the same time as the media-political system is being consolidated in Russia, the Russian language Internet is showing more and more life and developing increasingly complex forms" (Zassoursky, 2004:184). Media consolidation and government limitations on the Internet are not unique to Russia, of course—the U.S. Communications Decency Act of 1996 being a case in point—but the different backgrounds lead to different interpretations. In the United States, civil liberty concerns are preventative, ensuring that freedoms protected in our constitution are not threatened—while for many Russians, the issue of freedom of speech is imbued with the significance of a repressive past.

Zassoursky and other "Optimists" see the Internet as essential to the survival of media and information diversity. For instance, through online access to foreign media—both directly and in translation (as on the site http://www.inopressa.ru)—"any Russian user of the Internet can check the information appearing in the national media and weigh the situation from a different point of view" (Zassoursky, 2004:166). And though it is questionable how many people actually take advantage of these sources, it is important that they are available at all. "Against the backdrop of the reconstruction of the image of Great Russia, an image that dominates the symbolic field," he writes, "it is . . . sufficient that the Net allows communities and reference groups to exist outside the boundaries of political discourse and to work out their own cultural codes." And while it is perhaps "required more by a vocal minority than by a silent majority," he continues, it is nevertheless important, since "it leaves a window of freedom in the communications system, and does not allow the majority to drive dissidents into a 'spiral of silence'" (Zassoursky, 2004:184). It is interesting to note, in light of the LiveJournal licensing agreement discussed below, that Zassoursky was, until recently, a marketing director at SUP Fabrik.

Pessimists counter that the Internet, like any other social or cultural domain before it, may be easily taken over by elites and submitted to their interests. This group usually sees political and commercial forces as working together and is highly concerned with the vulnerability of the Internet to state control. For them, as for many Russian intellectuals, China serves as both an economically inspiring and politically frightening case in point.

As we hope to demonstrate below, these two cultural and political themes underlie and intertwine the complex and passionate discussions surrounding the shifting "ownership" and identity of the Russian LiveJournal community.

The Internet and Transnationalism—Virtual Reunification?

The Internet emerged at the same time as the Soviet Union was falling apart—combining a communications revolution with a political one. Although it took another decade for the Internet to enter the mainstream, for Russian intellectuals worldwide, the "Internet"—as both concept and entity—has intense symbolic value, representing communicative freedom not only within the country but with the rest of the world as well. Not surprisingly, censorship is a major trope.

The Russian language Internet has been, from its inception, a global phenomenon. Many early Internet resources within the Soviet Union were supported from abroad, both financially and organizationally (Andreyev, 1998). In the first years of the .su domain, for instance, traffic between Soviet and foreign users exceeded traffic within Russia (Ivanov, 2003). In contrast with the commercial and political impetus that spurred the development of the Internet in the West, the Russian language Internet evolved as almost a private initiative, led by a small elite, transnational group (Russian Cyberspace.org, 2006; Zassoursky, 2004:161). Many of these Russians were living abroad, "on the one hand deprived of their home country and on the other hand privileged by their access to the new media" (Schmidt, Teubener, & Zurawski, 2006:120).

In this context, Russia's reintegration into the global community—and reintegration with its own scattered diaspora—was marked, both symbolically and physically, by its connection to the World Wide Web. In particular, for millions of Russian dissidents and immigrants, the joint emergence of the Internet and dissolution of the Soviet Union portended a "virtual (re)unification" (Russian Cyberspace.org; Schmidt, Teubener, & Zurawski, 2006:120).

Today, the large numbers of Russian speakers living abroad have created a strong transnational community online, fulfilling, to a degree, the Internet's potential to reunify a splintered culture. At the same time, this "reunification" has also witnessed the emergence of an international and cosmopolitan subculture, as distinct from the domestic Russian one (Russian-Cyberspace.org). As one online study finds " . . . rather than leading towards a virtual reunification," the "Internet in Russia" and the "Russian Internet" form a "complex matrix of overlapping areas and distinct segments, producing constant fractions" (Schmidt, Teubener, & Zurawski, 2006:130). So, for example, in "The Changing Face of the RuNet," Bowles (2006) posits that the "differences between the RuNet and the rest of the Internet have gradually been dropping away, and . . . the RuNet is simply another backwater of the Internet, fenced in by a language barrier . . . but not essentially different" (30). In contrast, one of our LiveJournal survey respondents claims, "There are Russian Immigrant communities. And there are Russians who are living in Russia. And the culture surprisingly differs a lot."

Livejournal.Com—America's Most Popular Russian Blogging Site

The themes of communicative freedom, transnationalism, and identity discussed above are amply illustrated in the popularity of the American blogging and social networking site LiveJournal among Russian speakers worldwide.

Yandex, the main Russian language portal (www.yandex.ru)—similar to Yahoo and Google—consistently lists LiveJournal as the top Russian blogging site. Journalists and Internet scholars agree. In an early 2007 article on Russian blogging, Russian journalist Anna Arutunyan writes, "Walk into a typical Moscow newsroom and chances are good that half the people in your field of vision will be logged on to Zhivoi Zhurnal, the Russian incarnation of the American blog-hosting service LiveJournal" (Arutunyan, February 22, 2007).

LiveJournal was created in 1999 by a college student named Brad Fitzpatrick as a way to keep in touch with his friends. "Since then, it has grown into a user-supported, open-source service used worldwide." Today, LiveJournal is owned by Six Apart, a California-based blogging company (LiveJournal.com, 2007).

LJ is, essentially, both a blogging and social networking tool—users are connected to each other through a network of "friends" and "friends of." Journal entries are given security levels based, in part, on these friends lists, so that each entry can be private, or for "friends" only, or for a custom subgroup of friends, or completely public. In addition to creating journals, reading journals, and commenting on posts, LJ users can watch or join one of the numerous interest-based communities.

The LiveJournal statistics page shows that, since its inception, over 14 million accounts have been created. Of these, close to 2 million are "active in some way"—which means that they have been updated during a select time period. Users "from the Russian Federation" (and hence some, but not all "Russian users") make up the second largest group on LJ, according to the stats page, numbering over half a million.

It is important to note, as Kimmie Nguyen, product manager for LiveJournal, explains, that the stats page is a "pretty crude rollup" of the *public* information on a user's profile. It does not include those journals that make such demographic info private—which are many. In creating the stats, LJ works on the assumption that the information provided by users is true and accurate—"meaning that we don't expect many users are misleading others about their age, gender, and location." She notes that LJ has much more complex methods of gathering stats for proprietary, internal use (Nguyen, personal communication, May 30, 2007).

The Site's Russian Appeal

So how did an American social networking site, primarily popular with 18–22-year-olds, become the major Russian language blogging tool?

Eugene Gorny, academic and former editor at two major Russian online magazines (Zhurnal.ru and Russkij Zhurnal), has written extensively on this subject. According to his research, LiveJournal's popularity among the Russian-speaking community is based on the fortuitous collusion of key economic, technical, social, and cultural elements.

From a technical and economic perspective, free access and the existence of multilingual, Russian capabilities and interfaces allowed Russian language speakers to take advantage of the service, regardless of their economic status. In addition, from a sociopolitical angle, the fact that the site is "American," and, specifically, that user data have been stored outside of Russia has played into the theme of communicative freedom raised above, giving Russian users, both inside and outside the country, a sense of security. Ekaterina Alyabyeva underscores this liberating function of the Internet, comparing the RuNet with a Habermasian public sphere and pointing at its importance in a society where "private interests hamper the achievement of the public good and the emergence of indiscriminative rules of the game" (Alyabyeva, 2006:42). Sergey Kuznetsov, in his volume on history of RuNet, gives thrilling details on the role of LiveJournal during the hostage crisis in Moscow in 2002, when bloggers communicated with hostages via text messaging and published on LiveJournal information banned on all other media (Kuznetsov, 2004:378–381).

Finally, in terms of cultural elements, Gorny (2006) echoes the Slavophile perspective described earlier by positing that the social networking aspect of LiveJournal resonates with "the Russian tendency toward collectivist behavior" (77). It is interesting to contrast this perspective with American rhetoric surrounding social networking, which focuses on terms such as "collaboration" and "user participation," highlighting the many individuals who drive the process, rather than the collective.

Do Americans and Russians Use LJ Differently?

Gorny outlines the specifics of LiveJournal history and development among the Russian community, both in Russia and overseas. Initially, the site was what Gorny terms "a playground for intellectuals"—explored by a tightly knit, cosmopolitan group, discovering the Internet and LiveJournal's features and opportunities together. In the early years, a single page was available listing all Russian language LiveJournal users worldwide. This aggregate page eventually became too unwieldy, but echoes of it remain, particularly in the Yandex blogging portal (http://www.blogs.yandex.ru).

As the user base grew, more journalists, writers, political commentators, artists, musicians, and other "trendsetters" joined its ranks. Their blogs were more public than private, serving as online publications and articles on cultural and professional topics, with thousands of friend/readers not personally known by the authors.

"Since the beginning, [RLJ] has attracted a disproportionably high number of the 'Who's Who' in the informational and cultural space—journalists, writers, publishers, politicians, etc." writes online journalist Kirill Pankratov, adding " . . . once you spend time with the LJ, reading the same old Moscow Times . . . is the fastest way to put oneself to sleep" (Pankratov, 2005). By 2002, the media had dubbed the site "the most fashionable address on the web" (Gorny, 2006:75).

In a 2007 article on Russian blogging for *The Nation*, Arutunyan describes LJ founder Brad Fitzpatrick as "struck" by the "social magnitude of ZheZhe and the serious content of its journal entries." "What for Americans is an electronic diary accessible to a few chosen acquaintances became for Russians a platform for forging thousands of interconnected virtual friends," she writes (Arutunyan, 2007).

Part of the reason for this, according to Gorny, is that Russian users interpret the LiveJournal term "friend" differently. Gorny suggests that Russian speakers distinguish between the Russified "frend" (and its plural "frendii") and the Russian translation, "drug" and "druz'ja." LiveJournal "frendii" are not necessarily actual "friends" but often readers and other indirect connections. As a result, Gorny suggests, Russian users have more "friends" and "friends of" and hence a greater degree of interconnection between journals. And, though Russian LJ users are more likely to NOT personally know their friends, they also have "a tendency to de-virtualization," with user meetings often organized in Moscow and elsewhere (Gorny, 2006:76).

Nguyen sees the differences as more stylistic than cultural, explaining that on LJ, as in any social networking site, there are "power users" and "power communities/groups" that can have hundreds of friends and thousands of users. Nguyen reasons that demographics differ less by country than by the topic and type of community, so that, for instance, "celebrity communities tend to be younger than say communities around literary writing" (Nguyen, personal communication, May 30, 2007).

As the LiveJournal user base has grown, so have the number of private journals, and private/public journals, used more to keep in touch with friends than for online publication (Alyabyeva, 2006:40). Our pilot survey of Russian/American LiveJournal users found that many use their blogs as a way to keep in touch with friends both locally and abroad "without repeating the same story/event/occurrence over and over." For these young bloggers, half of their friends are in Russia, half are "cultural Russians" living in the United States. They write in a linguistic mishmash—one writes primarily in English (because he types it faster), but reads primarily Russian bloggers. Another explains how her language choice depends on mood and style "I post in English when I am very angry or rushed, and I use Russian when I have time to think the phrase through or am calm." They have chosen LiveJournal for the ability to instantly discuss things with people who are far away, for the capacity to keep private and public posts (unlike a standard blog) and to get a variety of opinions/feedback on posts/publications—not only from actual friends but from the broader community, including expert opinions. Their blogging focus is more social and personal than politically or publication oriented.

Despite the growing popularity of the site and the increased diversity of use and users, Gorny contends that the essential sense of community and interconnections remains. He quotes one user who describes "a communicative core in RLJ, consisting of a several thousands of journals that are tightly interwoven with each other. There remains a common communication environment in which the news spreads quickly and discussion about different political, literary or social issues can involve dozens of journals and hundreds of interested users" (Quoted in Gorny, 2006:75). Pankratov (2005) agrees—"Russian blogs are more concentrated on the LJ."

Today, the LiveJournal quota of 750 friends or 2,000 participants shown in poll results or a limit of 5,000 comments to one post—more than enough for Americans, who have an average of 10–20 friends—are still considered impositions for key Russian language users. The new Russian management of the sector, discussed in more detail below, anticipates lifting these bans—and profiting from the popularity of these blogs (SUP.com, 2007).

Controversy in the Blogosphere

"I Don't Want SUP, I Like BORSCH"

The issues relating to the parameters of the Russian language Internet explored above—especially identity and censorship—are clearly evident in the controversy surrounding a recent decision made by Six Apart/LiveJournal in regards to its large Russian community.

On October 19, 2006, LiveJournal/Six Apart announced a licensing deal for its "Russian segment" with the Russian media company SUP Fabrik, recently founded by American ex-pat and media/publishing entrepreneur Andrew Paulson and Russian financier Aleksandr Mamut.

The language surrounding the deal, on the part of Six Apart and LiveJournal in the United States, as well as SUP in Russia, indicates an unsurprising focus on the commercial interests and an ignorance, or willful avoidance, of the political and sociocultural implications. From the very beginning, Six Apart's press release blurs the distinction between the territorial and the transnational, between the Internet in Russia and the Russian language Internet. With a joint San Francisco and Moscow byline, the press release proclaims a "partnership to localize and improve the LiveJournal service for Russian speakers *worldwide*. Russia is the second largest territory for LiveJournal users, and LiveJournal is the most popular blogging service in Russia, with over 4.4 million monthly visitors to the site" (emphasis mine) (Six Apart.com, 2006).

This key distinction is neglected again on the pages of "Live Journal's Business Discussion Journal," where LJ Product Manager Kimmi Nguyen further explains the deal:

> We've licensed SUP, a Russian internet company, the right to use the LiveJournal brand. They'll be operating portions of the LiveJournal service for *our Russian userbase* (and only for those that want this support). This partnership is an effort to improve the speed, usability, and services that are offered *to our Russian community*. The only change that *Russian users* will see is an improvement to user experience, translation, and support. . . . *The Russian* users that have agreed to this support will still continue to be a part of LiveJournal, for as long as they want to be. SUP will be able to promote other products or services they develop to *the Russian community*, but they won't be able to disconnect anyone from LiveJournal or break apart the community. (emphasis mine) (Nguyen, 2006)

The posting emphasized that no journals were being transferred—that the deal only involved a change in "how the site will be supported." Nguyen assured readers that the option to "opt out of getting the new Russian features, and to continue to be supported by LiveJournal/Six Apart" would be available. In an email to the author, Nguyen clarified that "licensing the LJ brand means that Six Apart retains ultimate control over user's journals and the treatment of their accounts. Support entails getting support questions answered in Russian, getting pages on LJ properly translated into Russian, etc." (Nguyen, personal communication, May 30, 2007).

The announcement generated a furor in the Russian language blogosphere. Over the next few weeks, there were close to 700 responses posted on the LJ biz page—most overwhelmingly against the deal. "The Russian language blogosphere (commonly known as *ZheZhe*) is on fire" writes Khokholva in her blog for Global Voices Online, " . . . all because LiveJournal.com's owner *Six Apart* has decided to team up with the Russian Internet company *Sup*, founded this year by Aleksandr Mamut, a Russian 'oligarch,' and an American entrepreneur" (Khokholva, 2006).

Pessimists versus Optimists—The Role of Commercial Interests

From the perspective of LiveJournal, the deal is straightforward and commercial, a "win win situation" intended primarily for the benefit of Russian speakers in Russia and to tap Russia's expanding online community. "The goals of the agreement," writes Nguyen in a recent email to the author, are "1) to provide better support and services to Russian-speaking users and 2) to provide better local support and services to people in Russia (voiceposts, SMS, etc.). Users who don't opt out will get services that we can't easily offer to users in Russia such as SMS." As LJ founder Brad Fitzpatrick spells out in his personal LJ blog, "LJ promotes SUP, so SUP can be popular and make money later. SUP promotes LJ, so LJ's both better in Russia and can also make money there" (Fitzpatrick, November 1, 2006).

However, this commercially and nationally centered approach ignores the specific historical and sociopolitical issues of the Russian community, especially those outside of Russia—for whom memories of a repressive Soviet past intensify concerns with online freedoms and civil liberties. This group falls squarely into the "Pessimists" category mentioned above, conflating commercial and political interests in an anxiety over state control. Bowles writes that censorship concerns are "characteristic of the Russian Internet community, which fears the wider repressive ambitions of the Putin government. . . . With regard to the RuNet, the experience of the Soviet past is significant, and this might be seen as the stimulation for strongly articulated concerns about control" (2006:24).

Khokholva points out in her posting on Global Voices that "assurances from managers of Six Apart and SUP have left many unconvinced and still concerned whether the Russian security services would gain access to their personal information and whether the new Abuse Team would carry out ruthless *purges*" (2006)

(Emphasis mine, to highlight the use of the Soviet-era term "purges).")

> "People are worried about possibilities," explains one user. "The whole thing is inter-
> preted as getting 'a feet in the door' and expanding 'services' later . . . It is politics and
> it is the same everywhere—nobody is worried about Bush imposing a martial law over
> whole US and torturing all opponents, but the fear is that the next . . . president is now
> more likely to do that. . . . So far SixApart and SUP have made the abuses seem more
> likely. You have to address the perception, and not the math/deal terms. . . ." "LiveJournal
> is used as a political tool in Russia," he adds, "and the idea of a Russian company con-
> trolling even a tiny part of it is disturbing." (Krotty, 2006)

The issue played out in various media outlets, traditional and online, in Russia
and globally. In an editorial about the deal that appeared in the *International Herald
Tribune* only days after the announcement, journalist Evgeny Morozov (2006)
articulated the "Pessimist" fear that the media and political elites are united and that
the independent voices on the Internet are extremely vulnerable to state control:
"The consolidation of Russian media in the hands of people and institutions affil-
iated with the Kremlin has been almost completed. But as independent media
were fighting for survival, many dissidents found asylum online. Banned from television, radio and many newspapers, they had no choice but to start blogging. . . ."
He then connects these general concerns with the specifics of the SUP deal:
"Therefore, when a two month-old Russian start-up with the funky name of SUP
('soup' in English) announced last week that it would take over the Cyrillic segment
of LiveJournal from its American parent, the Russian blogosphere exploded with
buzz. Plenty of speculation about the Kremlin's vicious plan to control and censor
the blogosphere flooded the Internet." He follows with an analysis of the players,
concluding that, "given SUP's roots and potential ideology, one can hardly expect
that the scope of discussions allowed on the Russian Internet will increase. . . . If
history is anything to judge by, the days of the Russian blogosphere buzzing with
critical opinions are numbered . . ." he ominously predicts.

The *International Herald Tribune* editorial was referenced often in the ensuing
discussions of the deal in various online media. Most often it was cited as an example of the pervasive opinion on the deal within Russia. In addressing concerns surrounding the SUP deal on his personal blog, LiveJournal founder Brad Fitzpatrick
first refers readers to Morozov's editorial for insight and then presents his own, more
pragmatic take. From Fitzpatrick's perspective, views such as Morozov's are misguided "conspiracy theories." Rather, Fitzpatrick's "insider" analysis focuses on the
financial motivations of the players involved. "The Russian Internet boom is over-
due. There's a lot of money to be made. . . . Andrew Paulson is a businessman. He

likes to make money. . . . Mamut, more than anything, likes to make money." In Fitzpatrick's optimistic opinion, the Internet is an unstoppable force for freedom of speech, regardless of the political factors involved. " . . . Lay on the conspiracy theories, but I don't care . . . it doesn't make any sense. I believe Mamut at the end of the day wants to make money, not shut down blogging as a favor to the Kremlin. Because shutting down blogging is futile and he'd realize that. . . ." Fitzpatrick's view reflects a typical American "follow da money" faith in the power of commercial interests over political repression: " . . . the more you know, the less interesting it'll all seem," he concludes (Fitzpatrick, 2006).

Ironically, the attitude of *The Exile*, a Moscow-based alternative online paper, is closer to Fitzpatrick's than Morozov's. In his piece for *The Exile*, "The Last Supper of Russian Blogs," journalist Pankratov complains that the *International Herald Tribune* editorial simply reproduces a popular and misguided media frame: "Putin's dictatorship destroys the last vestiges of media freedom in Russia. Now it is reaching its tentacles into the sacred blogosphere. . . . Poor Russians! They simply don't get that 'democracy' and 'freedom' thing. . . ." According to *The Exile* piece, the furor had "little to do with 'Putin's brutal dictatorship'" and was more about a mishandling of publicity and public relations on the part of SUP. His conclusion on the future of the "Russian Blogosphere" is far more ambiguous.

Rather, the piece implicates the "mainstream Western" media, over actual Russian politics, in the perception of modern Russia as "repressive": " . . . Perhaps nothing sinister will come out of the Sup.com deal, and the Russian blogosphere will continue to thrive," muses *The Exile*. "In that case, one can be sure you won't hear about it in the mainstream Western media. Anything but the bad news about Russia is likely to be censored more bluntly in fact than any heavy hand of Putin's Regime" (Pankratov, 2006).

It is important to note here that, concerns with online censorship do not involve only the Russian government. *The Exile* recalls the "First Blog War," circa 2005, where users of ZheZhe were also fighting censorship—this time at the hands of LiveJournal's American Abuse Team, which, on an anonymous tip, shut down a journal with an offensive phrase—and then a few more blogs which repeated the phrase in solidarity with the original "victim." The theme among bloggers then was not only a serious concern with freedom of speech but also a jab at "the hypocrisy of self-righteous Americans stifling the self-expression of those wild and crazy Russkies" (Arutunyan, 2005).

Identity and Labels

In addition to the concern with civil liberty themes such as privacy, censorship, and freedom of speech, the LiveJournal announcement raised the volatile issue of identity—particularly, the power play between self-identity and imposed labels. The LiveJournal biz announcement stipulated that the criteria used to determine that a journal is "Russian" are "a combination of if you write primarily in Cyrillic, have listed your location as a country from the former USSR, or use a Russian browser" (Nguyen, October 18, 2006). Based on these criteria, a number of users, many with strong feelings about Russia, saw themselves suddenly, forcibly, and incorrectly labeled as "Russian."

A number of languages, aside from Russian, use the Cyrillic alphabet, including Bulgarian, Moldavian, Serbian, and Ukrainian. In addition, many users in "countries from the former USSR"—such as Ukrainians, for example, do not consider themselves "Russian" and, most importantly, do not *want* to be provided with a Russian-based service. And finally, there is the large Russian diaspora, many with powerfully negative memories of the Soviet Union, and with a deep-seated distrust of anything associated with Russian commercial or political interests. These people specifically chose LiveJournal as "their online home" because of its American ownership, and their reaction to having any part of their LiveJournal data "transferred" to Russia is akin to having themselves symbolically deported. In Russian, there is a linguistic distinction that directly addresses this issue. The adjective "Russkij" means "Russian" in terms of culture, whereas the adjective "Rossijskij" means "Russian" in terms of nation-state. So, for example, a member of the large Russian diaspora can understand herself as passionately "Russkij" (culturally Russian) while at the same time, being strongly anti-"Rossijskii" (Russian government, authorities, etc.).

Comments from disgruntled LJ users illustrate these points *(all emphases mine)*:

> "I am US customer and would like my relationship with LiveJournal to stay on this side of the Atlantic without me beggin somebody in Russia to let me use the service from the US company I have signed up with. . . . *I resent being classified as Russian citizen by the language I use.* This Internment Camp practice sound unethical if not illegal. If you sign the partnership with a company in Mexico, would you transfer all Spanish speaking users accounts there?" (Zaliva, October 20, 2006)

> "I prefer to have absolutely no contacts whatsoever with any establishment or institution based within the borders of *the former USSR* . . . I had already opted to LiveJournal, not any Russian-based blog, and the message was perfectly clear: I am staying away from whatever stays there." (Cathay Stray, October 19, 2006)

> " . . . For a lot of people who left, it feels as if they are being dragged back—without being asked. . . ." (Bagira, October 23, 2006)

> "It's not about contact with a culture, it's about subjection of certain infor-

mation streams to a certain STATE. . . . Those are two different cosmologies. *You'll never understand. You've never lived in the Soviet Union.*" (Myroslava, October 24, 2006)

"So here I am . . . USA IP, US English Browser, only most posts are in Russian. And Russian ads all over the place! Where's promised opt-out button! *Don't make me Russian just because I have friends in Russia!*" (MerigOO, February 7, 2007)

"I'm actually head of Lithuanian translation team and we're translating LJ to LITHUANIAN language, not RUSSIAN. If I use Lithuanian interface, the browser language is Lithuanian, *could you, please, give me any reason why should I be a part of Russian SUP support* and see links to LiveJournal.ru? Could you, please, answer me the question, why I'm automatically logged in this site Livejournal.ru? I would remind you, that Baltic states got independent in 1990, if you don't know this fact. And Lithuanian language does not use Cyrillic symbols." (deja voodoo, May 9, 2007)

From Nguyen's perspective, the controversy was relatively minor. In an email to the author, she explained that LJ news announcements often have many comments (often thousands). "There's *always* a set of users that react extremely to changes on LiveJournal. LiveJournal users are a passionate group of people, many of whom have strong opinions and varying viewpoints about anything and everything. If you take a look into any news announcement that we make (news.livejournal.com)—LJ'ers love the drama??? While we monitor the site to screen comments that we think would cause 'flame wars,' for the most part we let users freely express how they feel about the product, new features, partnerships, etc. . . . good or bad, we let them talk about it. And if an issue is valid, we do our best to address the needs of our users" (Nguyen, personal communication, May 30, 2007).

To that end, Nguyen responded to the slew of comments immediately. Her postings strove to reassure users on the safety of their personal information but played down the identity/labeling concerns:

"There seems to be a misunderstanding here. No one is being transferred or moved until they AGREE to it. And even if you decide you want to be supported by SUP, LiveJournal still has ownership of your journals. SUP is only licensing the LJ brand." "Nothing happens until you agree," she continues. "Users will be presented with the option to opt out before anything happens. . . . The choice of choosing SUP will be provided to you. If you don't want to, select NO and you won't be asked again. Specifically, she states' determining if your journal is Russian is only to tag users who may be interested in SUP. It does not mean you will be with SUP. That set of users will get a page where they can decide if they want SUP support. Select 'NO, not interested'—and you won't be asked again." (Nguyen, October 18, 2006)

The FAQ page on SUP's Web site, however, (http://www.sup.com/faq_en.html) presents the relationship differently. The questions and responses that it posts demonstrate a focused interest in the commercial potential of the Russian LiveJournal community and the intention to remain true to Six Apart rules, regulations, and values. Though they address many of the concerns raised by the Russian community overseas, the responses are not necessarily reassuring.

So, for example, the SUP FAQ page reveals an interest in expanding advertising, including removing, within the "Cyrillic sector," both the LiveJournal "750 friends per account" quota and LiveJournal ban on a user's ability to put ads on their blogs. "We will help all of interested Russian users to make sure that their blogging brings them not only fame and pleasure, but also commercial profits . . ." SUP explains.

At the same time, the company does not indicate that they make any distinction between Russian users in Russia and outside the country, explaining that SUP services will be offered to all "Cyrillic journals worldwide and journals posted from IP addresses in the ex-USSR." They concur that "the tag that defines the Cyrillic segment . . . identifies all journals that will be offered SUP services." However, rather than providing a specific "opt out" option, as Nguyen implies, the site explains that, while "SUP will never force anyone to use its services," there is also "no way to fully protect yourself from the offers. . . . If you don't want to use them, simply ignore them."

Similarly, SUP assures users that though they may have a license to the LJ trademark, "the property rights to the project, service and LiveJournal trademark remain in the ownership of Six Apart. Six Apart will continue to fully control compliance of SUP with all existing agreements with LJ users (such as Terms of Service and Privacy Policy)." They emphasize that the journals of Russian users will remain on the same database as the rest of LiveJournal and "all user content will be stored at the servers owned by Six Apart and physically located in California (USA)." Hence, journals already under U.S. jurisdiction before the agreement was signed will remain so, and "all law enforcement agencies will have to go to California with any requests to disclose non-public user information, which is kept in LiveJournal databases. The only valid reason for such a request to be served is a court ruling in effect."

At the same time, they also point out the important, and chilling, distinction between the physical location of a journal and the physical location of its author: "The U.S. legislation is only capable of protecting the journals and comments themselves, but not the authors of these journals who live outside of the U.S. If, for example, the prosecutor's office of the Altai region decides to prosecute some user from Barnaul based on posts that he wrote in LJ or some other blog or even in an email, the server's exterritoriality may not prevent the prosecutors from doing this. *This situation should be well understood by bloggers in all jurisdictions, where freedom of speech norms differ from those provisions found in the First Amendment to the U.S. Constitution.*" (emphasis mine)

The LiveJournal.ru Beta Site launched in May 2007, boasting over a million users and 63,000 communities.

Conclusion

The history of the evolution of the Russian language Internet is invariably a transnational one. At the same time, the initial popularity and dramatic growth of the transnational Russian language LiveJournal community were driven by specific socio-historical concerns. Similarly, the controversy surrounding the Six Apart/SUP licensing arrangement is rooted in geopolitics. This shift in understanding of what constitutes a "Russian user" and "the Russian community" online—corporate, personal, national, and transnational—depends on the perspectives, goals, and fears of the various parties involved, and their orientation on the Slavophile/Internationalization and Pessimist/Optimist axis.

In terms of the Six Apart/SUP licensing arrangement, the business players—both Russian and American—focus on the technological and commercial potential of the Russian language-based LiveJournal, neglecting the distinction between the transnational community and the local one (and ignoring completely all non-Russian Cyrillic users). In this context, it can be argued that, ironically, the business interests add a modern twist to the "Slavophile" approach, emphasizing national borders and needs and asserting that the universal technology of the Internet is best applied to the "Russian community" by a Russian company. They maintain an "Optimist" attitude in their disregard for the political implications of the "ownership" or management of the Russian language LiveJournal community. This attitude ignores or plays down the fears of those Russian émigré users who, subscribing to a "Pessimist" view, equate Russian commercial interests with political ones, and raise serious concerns about the future of their communicative and personal freedoms. Both groups view the Internet positively, with a key distinction in their emphasis on the relationship between commercial and political elites. It is interesting to note that the positive role of American business is implied and not questioned in this discussion.

The specifics of "the Russian Community" online illustrate the politics of identity construction on the Internet. As we have shown, the ability of social technologies to break the traditional bond between nation and culture and foster the emergence of transnational communities is tempered by the reassertion of traditional cultural and geopolitical understandings.

Addendum

On December 2, 2007, LiveJournal made a surprising new announcement—SUP Fabrik had gone beyond licensing the "Russian segment" of LiveJournal—they had purchased the entire company (www.news.livejournal.com). The Moscow-based company, the press release explains, would launch a new San Francisco–based company, LiveJournal Inc., to manage and operate LiveJournal content on a global scale.

For the various international users of LiveJournal, this transference of ownership is significant in different ways. In the first place, it effectively silences the passionate controversy over what constitutes a "Russian" user on LiveJournal—now, all LiveJournal blogs, regardless of their national or cultural affiliation, belong to SUP. For Russian users based in Russia, the sale is irrelevant, as their journals had already been licensed to SUP in 2006. In fact, the deal affects American users the most, because they are the ones to now experience the effects of multinational ownership firsthand.

The comments to the December 2nd announcement reached their 5,000-post maximum within two weeks. The responses seemed to combine stunned surprise with cynicism and cautious optimism. The main issue was the direction that this new company would take—while some users were concerned with how the deal would affect the privacy and autonomy of their journals, others were more interested in features, fearful that LiveJournal would lose its uniqueness and begin to imitate Facebook or MySpace. Similarly, though some were wary about this unknown company and the negative reaction it had garnered from Russian users, others were hopeful that SUP would live up to its promises and be more responsive to their needs than the previous owners, Six Apart.

In many ways, the story of LiveJournal, an American content producer purchased by an overseas media company, echoes that of other multinational media buyouts, with similar issues surrounding freedom of expression and the ownership of cultural production. Certain key elements make this scenario unique, however. The industry in this case is the Internet—not the music or film industry. LiveJournal Inc is one of the first examples of a foreign media company purchasing an American social networking site. And the purchasing company is not European, or Japanese—but Russian, with all the associated baggage of the Cold War. Most significant about this story, in my opinion, is how well it illustrates the complexities of globalization—playing out, in vivid detail, how the shifting tensions between transnationalism and traditional geopolitics complicate any simple correlations between nation and culture, identity and place.

References

Alyabyeva, E. (2006). O Socialnih funktsiyah blogov v sovremennoy Rossiyi. In V. Volokhonsky, Y. Zaitseva, M. Sokolov (Eds.), *Lichnost i Mezhlichnostnoye Vzaimodeistviye v Seti Internet. Blogi: Novaya Realnost*. St. Petersburg: St. Petersburg University Press.

Andreyev, A. (1998). *Fido, Usenet i Wce-Wce-Wce*. Retrieved September 2007, from http://www.fuga.ru/lexa/wse.htm

Arutunyan, A. (2007, July 1). *First blog war*. Retrieved June 2007, from *The Exile* Web site: http://www.exile.ru/2005-July-01.

Arutunyan, A. (2007, February 22). *In Russia's blogosphere, anything goes*. Retrieved June 2007, from *The Nation* Web site: http://www.thenation.com/doc/20070312/arutunyan

Bowles, A. (2006). The Changing face of the RuNet. In H. Schmidt, K. Teubener, & N. Konradova (Eds.), *Control + Shift. Public and private usages of the Russian internet*. Norderstedt: Books on Demand. Retrieved April 2007, from http://www.ruhr-uni-bochum.de/russ-cyb/library/texts/en/control_shift/Bowles.pdf

Ethnologue.com. (2007). *Russian—A language of Russia (Europe)*. Retrieved July 2007, from http://www.ethnologue.com/show_language.asp?code=rus

Fitzpatrick, B. (2006, November 1). LiveJournal, SUP, Russia links. Retrieved June 2007, from *Brad's Life* Web site: http://brad.livejournal.com/2261770.html

Gorny, E. (2006). Russian LiveJournal: The Impact of cultural identity on the development of a virtual community. In H. Schmidt, K. Teubener, & N. Konradova (Eds.), *Control + Shift. Public and Private Usages of the Russian Internet*. Norderstedt: Books on Demand 2006. Retrieved May 2007, from http://www.ruhr-uni-bochum.de/russ-cyb/library/texts/en/control_shift/Gorny_LiveJournal.pdf

Ivanov, D. (2003). *Istoria Russkogo Interneta. Khronologia*. Retrieved September 2007, from http://www.nethistory.ru/chronology

Khokholva, V. (2006, October 21). *Russia: The second blog war*. Retrieved June 2007, from *Global Voices Online* Web site: http://www.globalvoicesonline.org/2006/10/21/russia-the-second-blog-war/#more-16598

Koltsova, O. (2007). Media, state, and responses to globalization in post-communist Russia. In P. Chakravartty & Y. Zhao (Eds.), *Global communications: Toward a transcultural political economy* (pp. 51–74). Lanham, MD: Rowman & Littlefield.

Krotty (2006, November 1). *OK, lets do the math (Russian way)* Comment on *Brad's Life* Blog. Retrieved June 2007, from http://brad.livejournal.com/2261770.html

Kuznetsov, S. (2004). *Oschupivaya Slona: Zametki po Istorii Russkogo Internet*. Novoye Literaturnoye Obozreniye.

LiveJournal.com (2007). *FAQ—About live journal*. Retrieved April 2007, from http://www.livejournal.com/support/faq.bml

Morozov, E. (2006, October 25). Meanwhile, Russia's last refuge, the blogosphere. *International Herald Tribune*. Retrieved June 2007, from http://www.iht.com/articles/2006/10/25/opinion/edmorozov.php

Nguyen, K. (2006, October 18). LJ and SUP. *Live journal biz*. Retrieved June 2007, from http://www.livejournal.com/users/lj_biz/239637.html

Pankratov, K. (2005, July 1). *Censor this! Eye on blogs*. Retrieved June 2007, from *The Exile* Web site: http://www.exile.ru/2005-July-01/censor_this_.html

Pankratov, K. (2006, November 3). *The last supper of Russian blogs.* Retrieved June 2007, from *The Exile* Web Site: http://www.exile.ru/2006-November-03/the_last_supper_of_russian_blogs.html

Pilkington H., Omel'chenko E., Flynn M., Bliudina U., Strakova E. (Eds.). (2002). *Looking West? Cultural globalization and Russian youth cultures.* Pennsylvania: Pennsylvania State University Press.

Russian Cyberspace.org (2006). *Virtual (Re)Unification? An investigation into cultural identity performances on the Russian Internet.* Retrieved July 2007, from http://www.ruhr-uni-bochum.de/russ-cyb/project/en/project.htm

Schmidt, H., & Teubener, K. (2006). "Our RuNet"? Cultural identity and media usage. In H. Schmidt, K. Teubener, & N. Konradova (Eds.), *Control + Shift. Public and private usages of the Russian internet* Schmidt, Teubener, Konradova. Eds. Norderstedt: Books on Demand. Retrieved April 2007, from http://www.ruhr-uni-bochum.de/russ-cyb/library/texts/en/control_shift/Schmidt_Teubener_Identity.pdf

Schmidt, H., Teubener, K., & Zurawski N. (2006). Virtual (Re) unification? Diasporic cultures on the Russian internet. In H. Schmidt, K. Teubener, & N. Konradova (Eds.), *Control + Shift. Public and private usages of the Russian internet.* Norderstedt: Books on Demand. Retrieved April 2007, from http://www.ruhr-uni-bochum.de/russ-cyb/library/texts/en/control_shift/Schmidt_Teubener_Zurawski.pdf

Six Apart.com (2006, October 19). *Blogging leader Six Apart and Russian online media company SUP partner around LiveJournal.* Retrieved June 2007, from http://www.sixapart.com/about/press/2006/10/blogging_leader.html

SUP.com (2006). *Frequently Asked Questions* Retrieved April 2007, from http://www.sup.com/faq_en.html

Zassoursky, I. (2004). *Media and power in post Soviet Russia.* Armonk, NY: M. E. Sharpe.

CHAPTER SIX

Mapping THE Australian Political Blogosphere

AXEL BRUNS AND DEBRA ADAMS

The role of blogs and bloggers in the political process has received a great deal of attention in recent years, perhaps especially so in the context of the rise and fall of the first mainstream blogger-candidate for U.S. president, Howard Dean, and the subsequent scramble of other political players in the United States to establish blogs of their (or more frequently, their staffers') own. U.S.-based blogs have also been seen as important in driving political issues from the demise of Trent Lott to the growing opposition to the Iraq war. The role of blogs in the political spheres of other countries is less well understood, however; indeed, for other Western countries it is often assumed that blogs will operate there in much the same way as they do in the United States, though perhaps somewhat lagging behind the leader because of delays in adopting the technology or achieving critical mass.

In order to move beyond such generic assumptions, in this chapter we explore the political blogosphere in Australia. As an English-speaking nation of comparable living standards, with similar culture, and (at the time of writing in late 2007) in a broadly comparable political situation (such as prolonged conservative rule, participation in the "Coalition of the Willing" that invaded Iraq, and the politicization of terrorism threats)[1], it may be assumed that Australia's political blogosphere would show some of the same characteristics as that of the United States—however, we have found that for historical, social, and cultural reasons Australia has developed a blogging accent of its own.

Some Background

As an affluent and well-educated nation, Australia's use of information technology is broadly comparable with that of other developed nations; at the same time, however, its use of Internet technology has historically lagged behind a number of its North American, Western European, and South East Asian counterparts, and it has only recently begun to close that gap. This is particularly notable in the area of broadband Internet: here, while the technology for high-speed home broadband services is certainly readily available, cost/speed ratios remain significantly less attractive to consumers than they are in other nations (see, e.g., OECD, 2006b).

A 2005–2006 Australian Bureau of Statistics (ABS) report on Household Use of Information Technology (HUIT) shows that 60 percent of all Australian households have Internet access; 28 percent of all Australian households have access to broadband (see Figure 6.1).

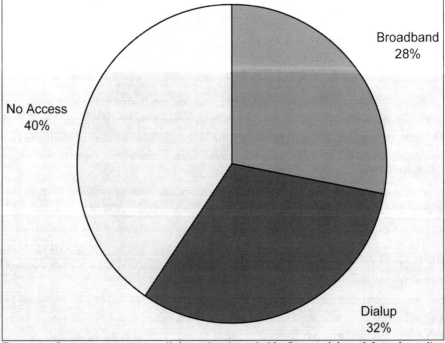

Figure 6.1. Internet access type—all Australian households. Source: Adapted from Australian Bureau of Statistics.

The comparatively high cost of broadband in Australia is due largely to historical factors: on the one hand, commercial cable broadband has as yet failed to recoup infrastructure costs and therefore remains expensive, while DSL broadband offerings use the existing domestic telephone network, which largely remains in the hands of recently privatized former monopoly telecommunications carrier Telstra (Clarke, 2004: 37), a company that has little incentive to lower the wholesale network access fees it charges its competitors. In addition, though such concerns apply in the handful of Australian metropolises that are home to the vast majority of the domestic population, telecommunications access in rural and remote areas of the country remains problematic overall, and obtaining reliable broadband access is a particular challenge (see Green & Bruns, 2008, for further details on such matters).

Such limited access to affordable high-speed, always-on Internet services affects Australian Net users' overall ability to engage in participatory online culture. Blogging, and political blogging perhaps especially so, operates commonly by drawing together and commenting on news articles, press releases, background information, and commentary from fellow bloggers, and the process of gathering such information and links is severely hindered by limited access and slow loading times; similarly, in the absence of an always-on Web experience, would-be bloggers may also comment less frequently on other blogs than they would otherwise prefer to do.

Access limitations and cost considerations may also skew participation in the Australian political blogosphere toward those strata of society that are better able to afford broadband access[2] or are able to blog and comment from their better connected places of work and study. This, then, would contribute toward an overrepresentation of relatively affluent and better-educated participants, and especially perhaps of white-collar workers and tertiary students (see Figure 6.2). It may also introduce an imbalance in favor of residents of Australia's major metropolises.

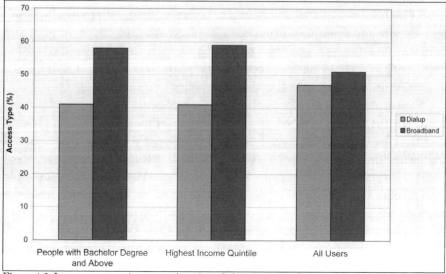

Figure 6.2. Internet connection type—by selected characteristics. Source: Adapted from Australian Bureau of Statistics 29.

Beyond the potential effects of the technological environments, social factors influencing Australians' degree of participation in blog-based communication are likely to be comparable to other Anglo-Saxon Western nations—technological literacy is generally well developed yet better distributed among the younger generations, the more affluent, and the better-educated; the gender gap for technological literacy has largely closed. As an English-speaking nation, Australia is also able to connect very directly with developments in the United States and the United Kingdom, avoiding the translation lag that may slow technology adoption in other developed and developing nations. Australians therefore have played a significant role in a variety of participatory culture projects in the past; Australian activist Matthew Arnison developed the software for the first *Indymedia* Web site that was used during the 1999 Seattle World Trade Organization protests (see, e.g., Bruns, 2005; Meikle, 2002).

Political Situation

Political blogging in Australia has emerged from such activist traditions at least in part; as we discuss below, a significant portion of the Australian political blogosphere continues to champion activist, progressive causes largely aligned with the left of the domestic and international political scene. This significant leftist leaning may date back to the advent of the Internet (and especially the World Wide Web) as a mainstream medium in the mid-1990s: since that time, Australian federal politics has been dominated by the conservative Coalition government (combining the Liberal and National parties and elected to power in 1996), with relatively consistent support also from most mainstream commercial media outlets in print and broadcast. New Internet media have provided a key space for the expression and discussion of both marginalized and oppositional viewpoints and the organization of activist events opposing federal government policy.[3]

Present-day Australian politics presents a somewhat more nuanced picture, however—the conservative federal government of Prime Minister John Howard now faces Labor premiers or chief ministers in all eight states and territories, and for the first time in some years, a popular leader of the federal opposition emerged in 2006 with a realistic chance of winning the general election in late November 2007.[4] Political opinion remains divided across a number of key issues: while the economy has performed strongly over the past decade, enabling the government to deliver nine consecutive budget surpluses since the 1998–1999 financial year, much of this success is attributed to the current resources boom and driven especially by the growing Chinese economy's demand for coal and other primary resources. Detractors therefore argue that economic success has been largely determined by environmental factors rather than government policy. The federal government also introduced a radical reform to industrial relations law in 2006, removing a significant amount of workers' protections against unfair dismissal and significantly reducing the ability of workers' unions to act on behalf of their members in collective contract negotiations. The workplace and industrial relations policy was initially marketed to the Australian public as "WorkChoices" at a cost of A$55 million, but has proven so widely unpopular with workers that the government recently rebranded the policy as the Workplace Relations System in the hope of gaining wider public acceptance. However, this action ignited new controversy for the government, which has been attacked for spending up to A$1 million of taxpayers' money per day on perceived party-political advertising in the lead-up to the federal election.

Further debate is driven by Australia's continued participation in the highly unpopular Iraq War and by the government's strong support for the Bush administration in that war and the wider "war on terror"; such support has also led to increasingly fraught relations between the government and domestic Muslim and civil rights organizations.

Overall, then, there is a growing sense of disenchantment with federal government policy; the government overall, and Prime Minister Howard specifically, are regarded as increasingly autocratic and out of touch with constituency sentiment. At 68, Howard's age and length of tenure as Prime Minister have also become an increasingly prominent issue; having intimated during the 2001 and 2004 election campaigns that if reelected he may stand aside mid-term in favor of Treasurer Peter Costello, he remains committed to fighting the 2007 election against 49-year-old opposition leader Kevin Rudd (with a renewed pledge to step aside for Costello during the next term, if reelected). Rudd's consistently favorable results in recent opinion polls are attributed largely to his presentation as a more responsive, younger, and more honest alternative to Howard (whom critics frequently portray as a cynical manipulator of public sentiment).

At the same time, large sections of Australia's mainstream commercial media remain firmly supportive of Howard. Australian news is dominated by a small number of organizations: between them, News Corporation and Fairfax Media operate the majority of national and state newspapers, including News Limited papers such as the *Courier-Mail* (Brisbane) and *The Western Australian* (Perth) that are the only local newspapers in these key state capitals. Television news is led by the nightly news bulletins of Network Seven and the Nine Network, while drive-time radio talk-back shows are a very significant factor in broadcast commentary on current affairs and are dominated by right-wing hosts such as Alan Jones and John Laws (Flew, 2003: 232; Pearson & Brand, 2001: 97). Against these, the radio, television, and online services of the public national Australian Broadcasting

Corporation (ABC) provide a more balanced view (but are frequently accused of Labor bias by Coalition politicians, even though independent studies have found little evidence of systemic bias; see, for example, Posetti (2001–2002) for a discussion of this controversy).

The Australian Political Blogosphere

Perceived pro-Howard bias in the commercial mainstream media in Australia may well act as a significant driver for participation in the Australian political blogosphere by left-of-center bloggers (as, indeed, may perceptions of leftist bias in the ABC drive right-wing blogging); such use of blogging as a corrective to apparent media shortcomings is consistent with phenomena in news blogging and other forms of collaborative online news production as they have been observed in other countries (see, e.g., Bruns, 2005; Bruns, 2006; Singer, 2006).

In order to examine more closely the make-up and structure of the Australian political blogosphere, we used the IssueCrawler Web site network mapping tool (www.issuecrawler.net) to explore the networks connecting Australian bloggers on political topics. Our methodology here follows in part the methodological approach explored in previous work by Bruns (2007); in addition, we undertook further in-depth case studies using different and complementary methodological approaches (outlined below). We analyzed the resultant network maps for notable patterns, as well as further examined the key sites that were identified in the crawl.

In the first place, this process produces momentary snapshots of current patterns of interlinkage between Australian blogs, weighted according to the frequency and reciprocity of interlinking. Since, for bloggers, linking to their peers and to other sources of information constitutes an act of constructing a distributed discussion across Web sites, such snapshots of link patterns indicate the sites around which political debate on specific issues currently center, and (when compared over time) can also indicate how the focus and locus of political debate shift in response to new news reports, press releases, political statements, and other published information. Abstracted from individual political issues, repeated over time, the process also points to the overall leaders of political discussion and commentary in the Australian Web, to their interconnection with the online presence of traditional Australian media organizations, parties, advocacy groups, and government bodies and may perhaps even indicate the traction that specific political issues and campaigns are able to gain in the wider electorate.

Finding 1: Considerable Left-Wing Inclination, Strong Polarization on Specific Issues

Two initial case studies conducted in February and March 2007 both point to a recurring finding in each of the subsequent explorations we have undertaken: participation in the Australian blogosphere currently appears to strongly favor the left wing of politics. Leftist political blogs exist in larger numbers and are more central to political discussions virtually across all issues we have examined. This is notable both in the case of the discussion of the fate of Australian-born Guantanamo Bay detainee David Hicks (examined in Bruns, 2007) and in the blog-based discussion of the controversy surrounding disgraced Western Australian political powerbroker Brian Burke during March 2007. While a leftist leaning for the former case—portrayed largely as an issue of human rights—is perhaps unsurprising, the latter case is more complex: Burke, a former Labor state premier of Western Australia, had been a successful lobbyist after his exit from politics but had been exposed for corrupt dealings and undue behind-the-scenes influence on state government ministers, leading to the resignations of several such state ministers. The federal government also attempted to use Burke's tenuous connections to federal Labor opposition leader Kevin Rudd to arrest Rudd's rising popularity, but such attempts backfired when newly revealed connections between Burke and conservative federal minister Ian Campbell forced Campbell's resignation instead. In the Burke case, therefore, both sides of politics could be expected to be equally represented in blog-based discussions of allegations and counterallegations, and Labor- as well as Coalition-aligned bloggers had considerable material for comment and debate.

In spite of the bipartisan nature of the Burke controversy, however, our results clearly show a significantly stronger representation of leftist and nonpartisan political bloggers in this case, and this trend continued in virtually all other cases examined. While the balance between left- and right-wing participation in the political blogosphere is certainly likely to depend crucially on the issues being debated, it is therefore unlikely that our choice of sample topics was the main factor determining our observation of a majority pro-Labor stance. Instead, these findings may be more indicative of a current political climate in which the ruling Coalition has found it difficult to gain political traction even from apparently fairly clear-cut issues favoring their side of politics; a pro-Labor bias in the blogosphere at present may merely reflect the pro-Labor bias in the wider electorate that polls have indicated since the beginning of the year (but may also point to a more fundamental leftist stance among the majority of Australian political bloggers, independent of current political events and determined instead, for example, by socioeconomic factors affecting online participation).

Further, we see some evidence for the gradual formation of polarized blog clusters around leading blogs on either side of politics: such clusters are made up in the main of blogs characterized by a shared political direction that exchange frequent links among one another but only link to oppositional views at a more infrequent rate. The Burke case suggests that such polarization may increase over time, as discussion moves away from the specific news reports which triggered initial debate, and returns to the topical heartland of the blog cluster (see Figures 6.3 and 6.4): repeating the same crawl one week later, we saw a marked contraction of blog-based discussion to a handful of key blogs, which now linked strongly only to blogs carrying similar ideological viewpoints within their own cluster. (Such polarization is also consistent with similar patterns observed internationally—see, for example, Adamic & Glance, 2007; Hargittai, 2005.)

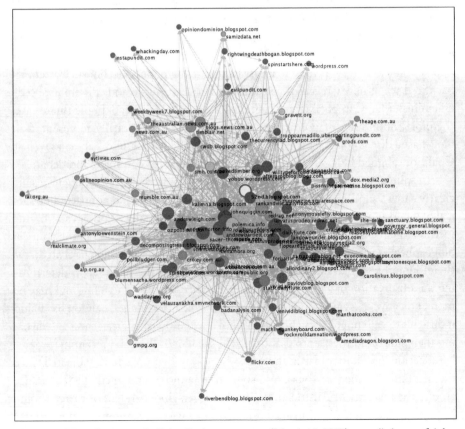

Figure 6.3. Blogs discussing the Brian Burke controversy (March 19, 2007)—small cluster of right-wing blogs in top left quadrant, remainder of core group is left-wing or nonpartisan.

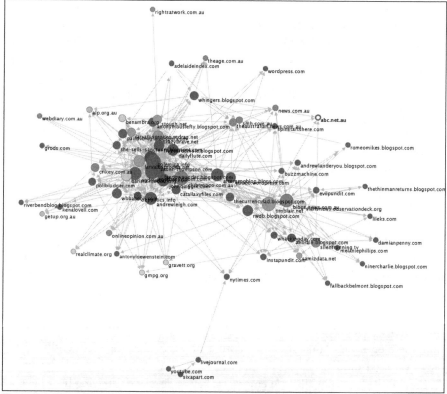

Figure 6.4. Blogs discussing Brian Burke controversy (March 25, 2007)—clear polarization between larger left-wing cluster (left) and smaller right-wing cluster (right), with few intermediaries.

It is possible for such contraction to be reversed again, too: this can take place, for example, when new news reports or other information restart or shift debate on an existing issue. We have observed such tendencies in longer-term studies of the government's "WorkChoices" industrial relations reforms, for example (discussed below). Here, the network of blog discussion of the WorkChoices legislation expanded and contracted parallel to increases and decreases in the level of coverage of the WorkChoices debate in the mainstream media.

Finding 2: Small Number of Core Nodes on Either Side

Such network contraction may well be a common phenomenon outside the Australian political blogosphere, too. Though initial blog posts are prompted by the emergence of news stories or other information and appear across a wide range of blogs, it seems appropriate to assume that subsequent postings on the same topic are more concerned with responding to commentary seen on fellow bloggers' sites; such commenting may take place increasingly on those sites themselves rather than at a distance through new posts on one's own blog. There is, in essence, an instinctive, low-key competition among bloggers to attract the focus of debate on any new topic to their own blog, and news and political bloggers often appear to do so by including in their first post on the topic a round-up of what other bloggers and the mainstream media have already said (following a process of gatewatching as described in Bruns, 2005). Especially as the network contracts, then, those sites that have performed best at this task and have attracted the most active subsequent discussion may emerge as the core nodes in the network though others diminish in importance.

Our studies indicate that, by and large, a small number of sites regularly perform this core function for their communities—this is likely to contribute to the gradual emergence and continued sustainment of an "A-list" of Australian political bloggers. Among the core leftist blogs in this context are *Road to Surfdom*, *Club Troppo*, and *Larvatus Prodeo*, as well as *John Quiggin* (especially on economic issues); also frequently connected to this cluster are the libertarian *Catallaxy Files* and the political philosophy blog *Sauer-Thompson*, as well as the site of the only notable federal politician to blog actively, Australian Democrats Senator Andrew Bartlett (the Australian Democrats are a minor, centrist-left party). The right is represented in the main by *Tim Blair*, as well as by the less consistent *RWDB* and *Yobbo*; in addition, the blog-style online opinion columns of political newspaper pundits in the News Limited and Fairfax press—especially of the controversial *Herald Sun* writer Andrew Bolt—occasionally also emerge as key nodes for right-wing blog debate.

Such sites may be seen in essence as "keepers of the flame" of political discussion for their respective ideological communities; they continue their coverage and debate of political events on given topics even at times when the overall volume of the main public discussion has decreased. This allows simmering issues to be examined and discussed in greater depth and from a wider range of perspectives. Other, more peripheral sites of group and individual bloggers appear to comment mainly when prompted to do so by coverage in the mainstream media; in the core sites of the political blog networks, on the other hand, discussion appears to be more detached from the day-to-day ebb and flow of political content in the conduits of industrial journalism and driven more strongly by sustained personal and group interests. Some bloggers in these sites have become leaders rather than followers, in other words, frequently posting new political ideas and analyses of their own rather than responding to and critiquing mainstream journalism's coverage.

Finding 3: Mainstream Media Act as Catalysts, but Not as Participants

In spite of their occasional interconnection with the blogosphere (which in addition to Andrew Bolt also involves the leftist *news.com.au* commentator Tim

Dunlop), mainstream media outlets and even some of the public intellectualism Web sites in Australia (such as *On Line Opinion*) are conspicuous in our results not quite by their absence but at least by their generally distant and marginal status. Even where they do offer commentary functionality or include outbound links in their content, such sites remain in use mainly as information sources rather than participating any further in the debate. This is hardly surprising for "traditional" sources of online news; it is somewhat unexpected, however, for the op-ed, blog-style content now offered, for example, by News Limited's *blogs.news.com.au*.

Such findings may indicate, then, that the pundit-bloggers employed by the major news organizations continue to operate under a different agenda from the mainstream of Australian political blogs: rather than participating in the wider, distributed political debate that takes place in the blogosphere beyond their sites, such pundit-bloggers are focused far more strongly on their local, on-site community of readers, in part perhaps also for commercial reasons of capturing a loyal readership rather than directing readers to alternative news and commentary sources. They appear to link out to other blogs less frequently and to pay limited attention to what blogs link to them; this also applies to their reader community, who in their comments are talking mainly among themselves rather than connecting their discussions to postings seen elsewhere. Such pundit-blogs, therefore, should be seen as constituting a separate environment in itself: not only separate from the core content of the Web sites established by the mainstream Australian news media but also separate from the mainstream of the Australian political blogosphere, they are a "safe" space that is populated perhaps by those users who enjoy their new-found ability to comment on columnist writings but have not yet made the leap to active participation in the uncontrolled blogosphere outside of the imprints of major news organizations.

Finding 4: Limited Reference to Outside Sources

The mainstream media are not alone in being sidelined by Australian political bloggers, however: the same is true also for the Web sites of government bodies, Australian politicians, and political parties, and other political and advocacy organizations. Indeed, our results appear to indicate that such references are more prominent early on in political discussions conducted in the Australian blogosphere and recede gradually into the background as the discursive network contracts and focuses more strongly on debates conducted across a handful of key nodes. Such tendencies were especially notable in a six-week study of debates around the "WorkChoices" industrial relations legislation: here, we conducted weekly crawls on the issue, starting from the first anniversary of the introduction of the "WorkChoices" legislation on March 27, 2007. During this time, we observed the peak of the debate followed by a gradual contraction and slowing of debate; this was followed by a government campaign against the ALP's proposed overhaul of the legislation if elected to power in 2007, which generally reintensified discussion of issues related to WorkChoices; beyond this, a gradual shift of focus toward wider economic issues took place in the lead-up to the presentation of the 2007–2008 federal budget on May 8, 2007 (an annual highlight of the Australian political calendar).

While media sources, as well as the sites of the Australian Labor Party's (ALP), public intellectual debate site *On Line Opinion*, and the Australian Council of Trade Unions' anti-WorkChoices campaign site *Your Rights at Work*, were present as marginal sites in the networks in week one, they disappeared over the following weeks as the debate slowed and contracted to the core sites and reappeared only briefly as the debate broadened again and shifted toward wider issues; notably, government sites (under the .gov.au top-level domain) were entirely absent throughout. This may point to a number of issues: on the one hand, the strongly partisan mainstream of participants in the Australian political blogosphere may see little reason to link to those political and informational sites that are already implicitly regarded as either allies or opponents; this applies for the larger leftist cluster of bloggers as much as it does for the smaller right-wing group (such tendencies may also point to a general well-developed understanding—and dismissal—by the electorate of the processes of political spin). On the other hand, the limited linking to such sources may also indicate an assumption that participants in the political blogosphere will already be familiar with current news reports and the political arguments presented by either side of politics; political blogging, in other words, is focused not on reporting political news but on discussing the implications of current political events, building on the information understood as already having been provided by journalists and politicians. This would seem to support a now well-established understanding of blogging as a discursive activity that complements mainstream journalistic coverage rather than replacing it altogether (it may well serve as a replacement for mainstream media *punditry*, however).

To the extent that the government, party, and advocacy group Web sites are provided in an attempt to bypass the reinterpretation and political spin of the mainstream news media and are addressed directly at citizens themselves, however, our findings appear to indicate that such efforts have as yet been largely unsuccessful—if bloggers *are* consulting such Web sites directly to gather information for their commentary of political events, they are at the very least not linking to them at any significant frequency. It appears that bloggers respond in their commentary not to the news releases of original political sources but to the reporting of such news in the mainstream media; this also means that the views provided by Australian political bloggers may therefore be commentary on the news media's coverage of politics as much as they are commentary on political events themselves.

Finding 5: News Media and Politicians Can Participate

Our investigations did point to a (small) number of cases where news media and politicians did actively and constructively engage with the Australian political blogosphere, however. Australian Democrats Senator Andrew Bartlett is the most visible (and possibly, the only) federal politician to have developed an active and persistent presence in the blogosphere, and his *The Bartlett Diaries* Web site regularly appears in a prominent position near the center of leftist blog discussion clusters we have encountered; on the site, Bartlett discusses freely and openly the key political issues of the day and engages, sometimes controversially, with commentaters; contrary to many other politicians' and media blogs around the world, he also frequently links to other sites of political debate. Similarly, *news.com.au* pundit-bloggers Andrew Bolt and Tim Dunlop occasionally do appear in more central positions in the network, again because both (Dunlop more so than Bolt) sometimes also link out to other bloggers and thereby further embed themselves in the distributed debate traced through such links.

To become a regular feature of the Australian political blogosphere, however, requires a significant amount of persistence, something that Bartlett has demonstrated, but others have not. For politicians as well as pundits, it means a notable shift away from a largely lecture-based style of delivery and toward a more discursive engagement with fellow bloggers. At the same time, such a repositioning also paves the way for increased peer criticism of one's own views and positions and may well be exploited by political enemies, a factor that likely discourages many other pundits and politicians from following suit. Bartlett, as a member of a relatively minor party struggling in recent years with high-profile defections and internal strife, may have more to gain from opening up to the electorate through blogging than he has to lose; the same is unlikely to be true of more prominent political leaders.

Further Analysis

In order to establish further detail about blogging and blog commenting practices, we now turn our attention from such broad examinations of the Australian political blogosphere to a series of more fine-grained studies. These focus on bloggers' coverage of specific issues and investigate more closely the processes of interaction and interlinking that take place on specific sites. Such studies depart from the overall methodology outlined above (and in Bruns, 2007) and add a process of in-depth manual coding of content and links to the IssueCrawler-based identification of overall blog networks.

Commenting on the Water Crisis

Our first case study in early 2007 examined blog-based discussion of the Australian water crisis. With many parts of Australia having experienced a prolonged, ten-year period of severe drought attributed to *El Niño* weather patterns and the overall impact of global warming, and many of its major cities as well as agricultural producers now under severe water use restrictions, the question of water management and the wider debate about responses to global warming have become an important issue in state and federal government politics; for the purposes of our research, the topic was also selected based on its long-term global currency and its connection to wider international debates about climate change. At the time of our examination, however, the water issue had been lying somewhat dormant in the mainstream media, while continuing to simmer in the blogosphere; it was only shortly after our study that the prime minister announced a new water initiative that generated broad discussions within both the traditional news media and the Australian political blogosphere. Our analysis describes the linking activity that occurred at various key political blogs and other relevant journalistic and institutional sites immediately before the prime minister's announcement; the resultant issue network map is shown in Figure 6.5.

Figure 6.5 displays a map with a number of key characteristics, most notably the presence of two clear (and clearly separate) clusters. One major cluster contains international organizations such as the World Bank, UNICEF, UNDP, IMF, OECD, and others, and is located in the top right quadrant. *The Guardian* and the Web site of the BBC are also present in this cluster but serve only as smaller nodes and have very few outlinks—one being from *BBC News Online* to *The Guardian*, indeed. There are a number of outliers: two U.S. National Oceanic and Atmospheric Administration sites and the U.S. Environmental Protection Agency (top) and the Australian Departments of Environment and Agriculture, the National Heritage Trust and National Resource Management (top left). Given a different set of seeds and parameters, this cluster could be studied further in isolation, and this would likely also identify other minor sites within the cluster; for our present purposes, however, such in-depth study is not required.

The cluster in the bottom left quadrant, by contrast, contains members of the Australian blogosphere. This cluster is made up of 14 sites: *Stoush, Crikey, Polemica, Sauer-Thompson, Larvatus Prodeo* (appearing under two different URLs), *Ampersand Duck, Anonymous Lefty, Pavlov's Cat, Club Troppo, LAN Downunder, Road to Surfdom, William Burroughs' Baboon,* and *John Quiggin.* In the primary analysis, clustering in this map indicates that the core Australian political blogs (most of them also identified as central to the blogosphere through our other crawls) are actively interlinked and occasionally link further outwards; few have their links reciprocat-

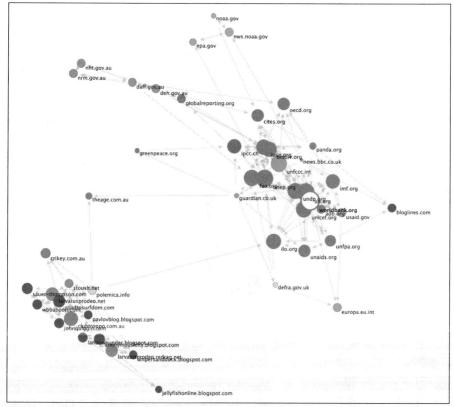

Figure 6.5. Water issue network, February 2007.

ed. Within both clusters, the size of nodes indicates their centrality—for example, consistent with our other findings, the left-of-center *Club Troppo*, *Larvatus Prodeo*, and *Road to Surfdom* are clearly core sites. Link direction further indicates that some of the outliers, such as political analysis and gossip site *Crikey* and Fairfax newspaper site *The Age*, are often linked to as key sources of information and opinion but rarely if ever link back.

We further conducted a detailed analysis of the outgoing links from four of the Australian blog sites listed above (see Figure 6.5). A search was conducted of each site for posts relevant to the issue around the date of the crawl. These posts were then examined for hyperlinks that were coded into four categories: issue-related blogs, news sites, reference sites, and personal (non–issue-related) blogs. From this, we found that 83 percent of the outlinks from the four sites selected for analysis were relevant to the issue (see Figure 6.6).

The outgoing hyperlinks at each of these blog sites are an indication of the participatory and collaborative nature of blog-based communication; they are key

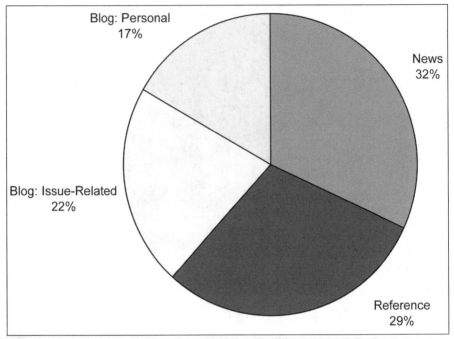

Figure 6.6. Type of hyperlink from selected sites discussing the water issue in the Australian blogosphere.

tools for the establishment and maintenance of a distributed discussion across sites. They also show that participants offer additional information to support or verify their viewpoints, share knowledge, and ensure the transparency of conversations—in this way all participants are able to receive instant feedback that allows them to measure their ideas and viewpoints against others.

Of the core sites identified here, then, the political commentary site *Crikey* contributed four stories about different aspects of the water issue. These covered topics such as government action over the water crisis, the cost of water, water consumption, and the effects of water minding in the suburbs. Such stories generated 12 comments, none of which contained hyperlinks to other sites of interest. The story posts themselves contained five hyperlinks—two linked to *Crikey*'s own information while two linked to traditional news media publications, *The Sydney Morning Herald* and *The Age,* and one is a signature/username link that points to a high-profile intellectual blog but not to an issue-related discussion. This indicates that, while occasionally described as a "blog," *Crikey* falls somewhat outside that category; while more blog-like in its content, in its engagement with other sites on the Web it operates more like a mainstream news site than like a blog, and its readers are similarly engaging in ways different from the commenters on core blogs in the cluster.

This is evident from the behavior of other core sites. For example, the *Sauer-Thompson* blog contains six posts about the issue and 32 comments. Topics of discussion are the shortage of water in South East Queensland, the notion of climate change as a myth, the question of whether the water shortage is a myth, irrigation rationalization, the sale of water to farms, and the city of Adelaide's dependency on the River Murray for water. There are 26 hyperlinks throughout the six stories and four in the comments. These divide into ten hyperlinks to traditional news media; seven hyperlinks to water authorities; four hyperlinks to issue-related blogs; and five hyperlinks to blogs that are not specifically related to this issue discussion.

Even more notably, *Larvatus Prodeo* generated a total of 92 comments through its three story posts. These stories are about water restrictions in Sydney and water minding in the suburbs—in particular, neighbors "dobbing"[5] on each other for wasting water. There are 7 hyperlinks within the stories, and 31 in the comments; 9 of the 31 hyperlinks are to traditional news media sites, 13 comment links are to authoritative sites, 11 link to issue-related blogs, and 5 are signature/username links that do not connect to issue-related discussions. While this sample of analysis is necessarily small, it nonetheless indicates that *Larvatus Prodeo* is the central point of discussion in this cluster of blogs at the time of the issue crawl (see Figure 6.7 for a summary of these findings).

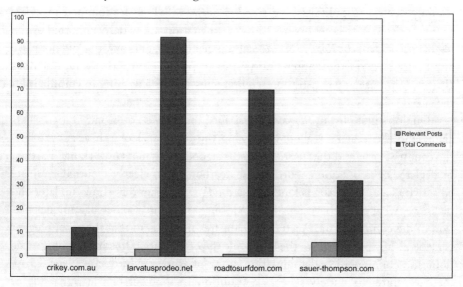

Figure 6.7. Number of water crisis-related posts and comments at selected sites.

It is important to note again here that this issue did not have significant currency in the mainstream media at this point of time yet evidently still managed to generate a strong level of reader involvement and feedback on several of the key sites of the Australian blogosphere. Given that the issue crawl was undertaken days *before* the prime minister released a new government water initiative that subsequently created wide media coverage, the results of our analysis show a commitment by these blogs to remain involved in longer-term conversations about issues that are of concern to the public, even while such topics are at least temporarily absent from the mainstream. It is also worth noting that coverage of the water issue itself was sometimes closely linked with wider themes of common interest to Australian political bloggers: so, for example, while *Road to Surfdom* did not post any water-related stories at the time of the crawl, during November 2006 the blog linked to a mainstream news media article about the water issue to point to what it perceived as political bias. This enabled commentaters to discuss both the water issue itself as well as its treatment in the commercial media.

From Blogger to Pundit

In pursuit of the forms of lively social, political, and cultural discussion common to blogs, News Limited recruited long-time Australian political blogger Tim Dunlop to create and run a blog-style section called *Blogocracy* on the news.com.au Web site (Dunlop, 2006). Dunlop is the creator and blogger-in-chief of the popular, left-leaning Australian political blog *Road to Surfdom*, which he founded in 2002. Dunlop's decision to accept News Limited's invitation was widely supported by the regular participants of *Road to Surfdom*; it was claimed that his new role did not place any limitations on the content he wishes to produce on either site, nor does *Blogocracy* impose any registration requirements on users wishing to comment on the site.

Dunlop's first post to *Blogocracy* described the purpose of his new blog as "less about journalism than it is about citizenship, the idea that all of us have a say in how the country is run and that participation is a good thing in its own right." Further, he declared, "I'm not doing journalism here, though it is clearly a journalism related activity," and "this blog is kind of like the forensic team that shows up later and picks over what others have already found." *Blogocracy* is in a somewhat unique position in the context of the Australian mainstream media, in that the chief blogger is ideologically positioned to the left, while the organization hosting the blog—the domestic component and foundation stone of Rupert Murdoch's News Corp. media empire—is widely seen as a traditional supporter of the right side of Australian politics. Further, in a marked departure from pundit "blogs" in many mainstream news media Web sites, *Blogocracy*, like most "normal" blogs, features a blogroll listing key bloggers in the political blogosphere—many of whom are also aligned with the political left. *Blogocracy* therefore can be understood as linking the features of traditional with those of new media, and as a point of connection for the politics of the left with those of the right.

In a further case study, we therefore sought to determine whether *Blogocracy* had retained the values and principles that characterize political blogs in Australia. We undertook a detailed analysis of the entire set of posts for a single week, May 21–25, 2007, to determine the kinds of topics discussed, and the types of outgoing links that appeared in posts and comments. Table 6.1 provides a summary.

Table 6.1. Summary of *blogocracy* activity, May 21–25, 2007 (internal outlinks point to material on the *news.com.au* site; external outlinks to URLs elsewhere on the wider Web)

	Date	Post Outlinks		Topics	Comments	Comment Outlinks	
		Internal	External			Internal	External
1	May 25	1	1	Jackie Kelly quits	41	1	1
2	May 25			Open Topic	143	1	21
3	May 25		8	Election Outcome Theories	36	1	4
4	May 25	1	3	Advertising Spending	30	1	2
5	May 24	2	4	Voter Intentions	44	2	1
6	May 24		3	Howard/Amnesty Int	20		
7	May 24	3	1	Industrial Relations	98	2	4
8	May 24	8	6	Industrial Relations	23	7	
9	May 23	2	2	Industrial Relations	66	4	
10	May 23	1	1	Leadership Change	86		2
11	May 23		1	Iraq War	95		8
12	May 22	1	1	Industrial Relations	76	4	5
13	May 22	1	2	Future Fund	34	1	2
14	May 22	1	2	Right-wing Writings	20		1
15	May 22	1	2	PM's Dining Room Refit	97	1	4
16	May 21		3	Industrial Relations	24		2
17	May 21	2		Election Strategies	52	3	1
18	May 21	1		David Hicks	59	1	3
19	May 21	2	1	Industrial Relations	48		1
	TOTAL	27	41		1,092	29	62

From this it is self-evident that *Blogocracy* is a very active site, with five or more posts added each day, on a variety of topics, and a relatively large number of comments for each post (57 per post, on average); *Blogocracy* is notably different in this from other sections of the *news.com.au* site, such as the *NewsBlog*, a section where readers are able to post comments that discuss the news of the day but which generated less than one post per day during the same timeframe, and featured no links to external resources. By contrast, our data for *Blogocracy* clearly show that its posts and comments contain a significant amount of both internal and external hyperlinks; such links point to background and additional information available both from official and mainstream media sources as well as from the wider blogosphere. Of particular interest here is the large number of hyperlinks to sites such as the *Sydney Morning Herald* and the Australian Broadcasting Corporation (ABC): the high rate of linkage to the *Sydney Morning Herald* by Dunlop and his commenters indicates the perception that issues covered at that mainstream media site are not covered by the publications of *news.com.au* itself; so, for example, a *Blogocracy* post by Dunlop on May 25, titled "Between Information and Party Promotion" linked to the *SMH* rather than to internal News Limited articles for additional information. (Indeed, a search of *news.com.au* was unable to find any reference to the topic of Dunlop's post—government spending on an advertising campaign promoting its preferred response to climate change.)

Other external links from *Blogocracy* cover a broad range of sites, including the Office of the Prime Minister, the Reserve Bank, and international news media, but almost half of the external links from Dunlop's own posts are to Australian political blog sites (Table 6.2). Overall, this shows that Dunlop has retained a freedom to generate and engage with conversations about issues that fall outside of News Limited's coverage or underlying ideological framework; indeed, he is in a position to connect the networks of conversation in the blogosphere with traditional news media networks.

In Dunlop's post and subsequent reader comments, blog sites account for approximately 30 percent of all external links. The most linked-to blogs include *Road to Surfdom* and *Peter Martin* (four times by Dunlop), *Polemica* (three times by Dunlop) and *Possum Comitatus* (once by Dunlop and twice in comments). Another 30 percent of external sites are commercial news sites—in fact, 22 of the 31 links to commercial news sites are to the Fairfax newspaper *Sydney Morning Herald*, while another three are to its sister site, the Melbourne-based *The Age*. The high rate of links to competing Fairfax publications is perhaps a reflection of where *Blogocracy* readers themselves prefer to find the news of the day—16 of the 22 links to the *Sydney Morning Herald* are found in the comments. This would further support our observation of a general leftist leaning in the Australian political blogosphere or at least is likely to indicate the predominant political preferences of Dunlop's audience,

Table 6.2. External sites linked to by type

Type of Site	Unique Sites	Outgoing Hyperlinks		
		Posts	Comments	Total
Alternate News	2		2	2
Blog	15	18	14	32
Charity	1		1	1
Commercial	3		3	3
Commercial News	7	11	20	31
Corporation	2		2	2
Government	3	7	6	13
Organization	4	1	3	4
Politician	2	1	1	2
Public News	1	3	7	10
Reference	1		1	1
University	1		1	1
Video Reference	1		1	1
TOTAL	43	41	62	103

as News Limited is perceived to be positioned clearly to the right of the center of Australian politics, where Fairfax is generally seen as taking a centrist or even slightly left-of-center approach. (At the same time, however, it must also be pointed out that outside of Fairfax newspapers and the commercial broadcasters, there essentially *are* no other mainstream commercial news organizations in Australia, creating a severely limited range of linking options for users wishing to point to the coverage of domestic news outside of News Limited.)

Government sites account for only 13 of the 103 external links; these links are mainly provided to source statistical information or to quote specific government policy documents or press releases. This further supports our general finding of the relative absence of such primary sources from blog-based political discussion in Australia. Finally, the next largest group of external links are to publicly funded news sites, or more specifically, to the online news resources provided by the Australian Broadcasting Corporation.

This case study provides only a snapshot of the range of topics and external links found at *Blogocracy* within a single, randomly selected week, of course. However, they do clearly match and support the overall tendencies that we have outlined in our general findings about the Australian political blogosphere above; this, then, would appear to indicate that Dunlop's *Blogocracy* site has been able to retain the features of a "traditional" political blog even while being hosted by a commercial, mainstream news organization. At the same time, *Blogocracy*'s relative absence as a core site of blog-based political discussion from the various general issue crawls we have conducted may indicate that Dunlop's move into the mainstream has both opened up a new community of readers and participants and failed to encourage existing users to make the move from *Road to Surfdom* to *Blogocracy*. This could also be read as a sign of deep distrust for News Limited's online and offline publications among the leftist mainstream of Australian political bloggers, in spite of the company's embrace of Dunlop as a staff blogger-pundit whose politics are significantly at odds with the prevailing political orientation of the corporation itself.[6]

Conclusions

The strong leftist tendencies and notable polarization of the Australian blogosphere may well be an indication that political blogs in Australia are at present mainly preaching to the already converted on either side of the political divide rather than playing a particularly active part in public deliberation on political issues and events. At the same time, however, it is also important to note that according to current political polling, the overall sentiment expressed in the blogosphere is broadly in tune with the prevailing mood of the wider electorate, which after 11 years appears to have become disillusioned with the Howard government and is expected to swing significantly toward Labor at the federal elections in late November 2007. Should a change of power occur at the federal level, it will be interesting to examine whether the political focus of the Australian blogosphere remains in step with wider public opinion even at a point where Labor's poll results may begin to fade again or whether the Australian political blogosphere is fundamentally and persistently pro-Labor (or at least pro-left wing) irrespective of the prevailing political climate in the country. It will be important to examine whether the election of a Labor government (and a possible shift of news media support to the election winner) generate a converse trend of increased blogging activity by conservatives, and a decrease of political blogging by Labor supporters; this, then, would point to a use of political blogging as a tool mainly for the expression of views in opposition to the government of the day and to the mainstream media sentiment.

Clearly, such developments are as yet difficult to predict and are further influenced also by the changing demographics of active online participation as access to the Australian Internet in general, and high-speed broadband more specifically, becomes more affordable and citizen literacy in participatory online media improves. Our present observations therefore mark only the beginning of what must be an ongoing project of research into political activity in Australia using participatory online media and must be extended through further research that takes into account the changing political landscape in this country. What is already evident, however, is that especially in the context of a highly concentrated and ideologically homogeneous mainstream of (commercial) Australian news media, political blogs in this country play a crucial role of adding a more diverse and multifaceted range of perspectives.

Acknowledgment

ABS data in this chapter have been used with permission from the Australian Bureau of Statistics (http://www.abs.gov.au/).

Notes

1. Eleven years of conservative rule at the federal level in Australia under Prime Minister John Howard ended on November 24, 2007, with the election victory of the Australian Labor Party under Kevin Rudd. This chapter was completed before this change of government could have a substantial impact on the Australian blogosphere, however, and documents the state of Australian political blogging during the last phase of the Howard government. For this reason, we continue to refer to the Howard government and its actions in present tense.

2. The ABS report shows a higher rate of broadband connection where households have higher income, higher education, and have members aged 15–24 years (ABS, 2006:29).

3. The potential for Internet media to attempt to balance the power and dominance of traditional media is of special importance in Australia, given its high levels of ownership concentration in such traditional media. Although many democratic societies around the world are experiencing similar problems, Australia has one of the highest levels of media ownership concentration in the world.

4. Voter polling showed the Australian Labor Party (ALP), led by Opposition Leader Kevin Rudd, to have a clear lead in citizen voting intentions over the incumbent government, led by Prime Minister John Howard (see Roy Morgan Research, 2007). The ALP subsequently won the November 2007 elections with a convincing majority.

5. An Australian slang term for reporting bad behavior, similar to "snitching."

6. As it turns out, such distrust may have been justified, indeed: News Limited's promises of editorial freedom for Dunlop were shown to be unreliable by an incident on July 12, 2007,

in which a *Blogocracy* post sharply critical of an editorial in News Limited's flagship paper *The Australian* (which had personally attacked prominent bloggers for disagreeing with the paper's political columnists in their interpretation of political polling) was removed from the news.com.au site without explanation (see Bruns, 2008; Bruns, Wilson, & Saunders, 2007).

References

Adamic, L., & Glance, N. (2007, March 16). *The political blogosphere and the 2004 U.S. election: Divided they blog.* WWW 2005 2nd Annual Workshop on the Weblogging Ecosystem: Aggregation, Analysis and Dynamics, Chiba, Japan. Retrieved May 10, 2005, from http://www.blogpulse.com/papers/2005/AdamicGlanceBlogWWW.pdf

Australian Bureau of Statistics (ABS) (2006, December 18). *Household use of information technology.* Retrieved May 31, 2007, from http://www.abs.gov.au/Ausstats/abs@.nsf/0/acc2d18cc958bc7bca2568a9001393ae?OpenDocument

Bruns, A. (2005). *Gatewatching: Collaborative online news production.* New York: Peter Lang.

Bruns, A. (2006). The practice of news blogging. In A. Bruns & J. Jacobs (Eds.), *Uses of Blogs* (pp. 11–22). New York: Peter Lang.

Bruns, A. (2007). Methodologies for mapping the political blogosphere. *First Monday, 12*(5). Retrieved May 14, 2007, from http://firstmonday.org/issues/issue12_5/bruns/index.html

Bruns, A. (2008). Life beyond the public sphere: Towards a networked model for political deliberation. *Information Polity 13*(1–2), pp. 65–79.

Bruns, A, Wilson, J., & Saunders, B. (2007). *When audiences attack: Lessons from the Australian poll wars.* Leeds: Centre for Digital Citizenship.

Clarke, R. (2004). The emergence of the internet in Australia: From researchers' tool to public infrastructure. In G. Goggin (Ed.), *Virtual nation: The internet in Australia* (pp. 30–44). Sydney: UNSW.

Dunlop, T. (2006, November 6). *Some call him Tim.* Retrieved May 31, 2007, from *Blogocracy* Web site: http://blogs.news.com.au/news/blogocracy/index.php/news/comments/some_call_him_tim/

Flew, T. (2003). A medium for mateship: Commercial talkback radio in Australia. In A. Crisell (Ed.), *More than a music box* (pp. 229–246). Oxford: Bergham.

Green, L., & Bruns, A. (2008) .au: Australia. In F. Librero & P. Arinto (Eds.), *Digital review of Asia Pacific 2007/8* (pp. 92–101). Montreal: Orbicom.

Hargittai, E. (2005, May 25). *Cross-ideological conversations among bloggers.* Retrieved March 16, 2007, from *Crooked Timber* Web site: http://crookedtimber.org/2005/05/25/cross-ideological-conversations-among-bloggers/

Meikle, G. (2002). *Future active: Media activism and the internet.* New York: Routledge.

OECD (2006a). *Broadband statistics to December 2006.* Retrieved March 14, 2007, from http://www.oecd.org/sti/ict/broadband

OECD (2006b, April 7). Working party on telecommunication and information services policies. *Multiple Play: Pricing and Policy Trends.* Retrieved May 14, 2007, from http://www.oecd.org/dataoecd/47/32/36546318.pdf

Pearson, M., & Brand, J. (2001). *Sources of news and current affairs.* Sydney: Australian Broadcasting Authority.

Posetti, J. (2001–02). The politics of bias at the ABC. *Australian Studies in Journalism* (10–11), 3–32.

Roy Morgan Research (2007, May 20). *Morgan poll trends*. Retrieved May 30, 2007, from http://www.roymorgan.com/news/polls/trends.cfm/

Singer, J. B. (2006). Journalists and news bloggers: Complements, contradictions, and challenges. In A. Bruns & J. Jacobs (Eds.), *Uses of blogs* (pp. 23–32). New York: Peter Lang.

Sydney Morning Herald (2007, May 15). *Government's budget bounce fails to materialise*. Retrieved May 15, 2007, from http://www.smh.com.au/news/national/governments-budget-bounce-fails-to-materialise/2007/05/14/1178995082856.html

CHAPTER SEVEN

Blogs AS Public Property Media

Defining the Roles and Assessing the Influence of Political Blogging in Israel

CARMEL L. VAISMAN

Recently, researchers have been engaged in mapping emerging relationships between the mass media, the political sphere, and the blogosphere (Adamic & Glance, 2005; Johnson, 2006; Lowrey, 2006; Robinson, 2006). Among these efforts, studies of American blogs engaged in random acts of journalism have been conceptualized as a form of participatory journalism (Caroll, 2004; Gallo, 2004; Gill, 2004). Indeed, this view was supported by a report that 34 percent of American bloggers see their blog as a form of journalism rather than as a personal diary (Pew Internet & American Life Project, 2006).

An additional approach seeks to understand the influence of American blogs in relation to their impact on national mass media practices and politicians. For example, anecdotal evidence has been used to assess the role of blogs in the resignations of Dan Rather of CBS and Eason Jordan of CNN (Johnson, 2006; Lowrey, 2006) or uncovering the Clinton-Monica Lewinsky affair and the resignation of Senator Trent Lott (Adamic & Glance, 2005; Gill, 2004; Johnson, 2006), following facts uncovered by bloggers. Further, there is evidence of an increasing number of politicians who interact with their voters directly as bloggers and as blog readers (Adamic & Glance, 2005). Despite increasing blog readership, these analyses suggest that while such blog influence is due to successful interaction with national mainstream mass media, it has only a short-term impact.

Accordingly, a number of key questions remain open: What are the media roles and functions performed by blogs? Is it possible that blogs will achieve an independent media status?

With these studies and questions in mind, this chapter presents an analysis of political blogging in Israel. Such a case study of the roles and impact of blogs is insightful because Israeli political blogs have limited interaction with national mainstream mass media.

This chapter is informed by participant observations and anecdotal evidence from the Israeli blogosphere conducted by the author as an active participant-observer during the years 2003–2007. The case study presented analyzes three leading political blogs in the Hebrew-language blogosphere, along with observations about the small English-language Israeli-based blogosphere. Data collection included structured interviews with founders of these blogs as well as analysis of the blogging efforts undertaken by a few members of the Israeli parliament (the Knesset). My goal has been to examine and to assess the roles and the influence of these blogging efforts.

A Brief History of the Israeli Blogosphere

Israel was an early adopter and a hotbed of cutting-edge technologies and Internet applications, such as instant messaging (ICQ), Firewall (CheckPoint), disk-on-key technology and so forth. In 2006, 3.9 million Israeli residents were connected to the Internet (52% of the population); 72 percent of the Jewish households were connected, 94 percent of them by fast cable or ADSL connection (Koren-Diner, 2006). And yet, the evolution and popularization of the Hebrew blogosphere remain underdeveloped and consist of as few as 50,000 active blogs.[2] This may be one reason why blog-related issues have not yet been surveyed as a nationwide phenomenon. The blog phenomenon formally emerged in the Israeli Web-sphere in August 2001 with the creation of *Isra-blog*, a Hebrew blog-hosting Web site that remains the largest popular blog-hosting Web site in Israel today. Before 2001, Internet-savvy practitioners were already writing English-language blogs on international Web sites or opened scattered personal blogs in Hebrew on private domains. As a result, when *Isra-blog* was founded it was populated largely by adolescents.

In March 2003, the desire to function according to the American blogging model led a group of Net veterans to create *Notes*, a Hebrew blog-hosting Web site that one can join only when invited or reviewed and approved by the founders. The idea was to attract local elites from various fields to blog by creating a "lab experiment" of the desired genre. At present *Notes* hosts 182 blogs.

In 2005, the second popular Hebrew blog-hosting Web site, *Tapuz's Blogosphere*, evolved from active online communities and forums. During 2006, two bloggers voluntarily translated the Wordpress[3] template into Hebrew, enabling the popularization of independent blogs in Hebrew. This resulted in the creation of a third Hebrew blog-hosting Web site, *Blogli,* that is Wordpress-based. In addition, *Kipa* [skullcap], a Web portal for the online religious community hosts 30 blogs by writers invited from this community.

Isra-blog and *Tapuz's Blogosphere* currently contain 85 percent of the Hebrew blog corpus. In August 2006, *Isra-blog* reported 74 percent of its bloggers were under the age of 17 and *Tapuz's Blogosphere* reported 51 percent were under the age of 20. Recently, *The Marker* financial magazine created a social network including blogging options that is gradually attracting additional audiences from the business community to the blogosphere.

Most blogs were produced as personal diaries or fictional literature genres over the course of the first few years of the Israeli blogosphere. *Israelis in the Dark*, written by Ofer Landa—a self-proclaimed right-wing Orthodox Jew and computer engineer—was the first politically oriented blog. In February 2003, Landa sent a strident email to each of the 120 members of the Knesset blaming them personally for the disintegration of the Israeli social infrastructure. Landa posted the letter on his blog and called for other bloggers to take similar action.

Landa's initiative was met with cynicism that turned into enthusiasm when seven personal responses were received from leading Knesset members, including one phone call asking for Landa's policy suggestions. A few other bloggers showed support by emailing Landa's blog link to Knesset members who had not yet responded to the letter. As a result of this extraordinary initiative, one Knesset member, Avshalom Vilan of the Meretz Party (left-wing)[4] subscribed to Landa's blog. Landa created the first direct contact between bloggers and politicians, establishing a successful model for blogger activism that was later adopted by *Black Labor,* one of the blogs analyzed in this chapter. Ehud Yatom of Likud party (moderate right-wing), the first Knesset member who attempted to blog, was in fact inspired by this interaction.

Local popular aggregation indexes, *Politinet* and *Shama*, that reflect the daily agenda of the Hebrew blogosphere, list approximately 250 bloggers who often write about political and social issues within their personal blog. An examination of the popular topics addressed shows Hebrew bloggers relate mainly to local political, economical, and social issues. Much less attention is addressed to security or matters related to the Occupied Territories, topics covered extensively by national and international mass media. These submissions react to mass media stories and rarely present original ones.

The Hebrew blogosphere is a closed homogenous community. The majority of those leaving comments on blogs are fellow bloggers. The entire phenomenon has little resonance outside the blogosphere. Since online writing represents both writing and speech, it is hard to measure and debate if political posts on blogs are like mass media opinion columns or conversations with friends that preach to us, hoping to experience catharsis.

Landa's attempts to engage in direct blogger-politician dialogue were unique until last year, following the 2006 elections, when the Israeli blogosphere witnessed the creation of three blogs divorced from the personal journal genre whose sole purpose was to involve other bloggers in a continuous political agenda: *Peer Pressure*, written by two journalists, offered a counter reading of news and events that sought to expose the faults of the social and political system; *Black Labor*, written by 30 journalists and activists, sought to create an active online community of Labor Party voters; and *KnessetWatch* offered participants an opportunity to engage in participatory legislative critique.

Although any form of blogging could be conceptualized as political in the general sense, I restricted myself in this study to blogs that could be compared with American political blogs from a mass media perspective, while conforming to traditional definitions of online political participation (Norris, 2001): that is, they involve a degree of congregation, a particular contact with the political sphere or protest politics. *Peer Pressure* and *Black Labor* are considered by bloggers and computer and Internet media channels as the leading political blogs in Israel and are often cited as the only examples of serious blogging in the Hebrew language.

The Depolitization of the Israeli Blogosphere

Although Israel is a vibrant country offering various news-making opportunities, on both national and international levels, and despite the fact that a number of leading journalists and political activists blog extensively, blog-related news have appeared to date on the main news pages only in dysfunctional social contexts or as cases of personal voyeurism, as when bloggers commit suicide or stir the community by discussing their pedophilic or extremely racist inclinations. One explanation for this phenomenon is the fact that adolescent girls dominate the statistical profile of the Israeli blogger (Vaisman, 2006).

Since the majority of Israeli bloggers exist as individual youth discourse clusters that seldom address issues of the collective, they have little reason to interact with the national mass media, other than when the latter spotlights the activities of Israeli children and youth. In this regard, it is worth noting the claim that the statistical majority of blogs in the United States, too, is dominated by adolescent girls (Herring, Kouper et al., 2004), a fact that has not interfered with the attention directed to other blog genres.

On the other hand, some of the most well-known bloggers in the United States are also employed as professional journalists and are referred to as j-bloggers

(Robinson, 2006). Surely the acceptance of the blog as a journalism genre and some of its influence stems from the professional identity of these practitioners. Furthermore, in Israel, some personal journals have received media attention in entertainment sections, and various such bloggers have been recruited as journalists. Given this situation, ignoring political blogs is all the more evident. Such limited interaction between Israeli j-bloggers and political bloggers with national mainstream mass media has to be understood in the context of the Israeli media ecology.

Israel has its own model of media ecology, one that is neither authoritarian nor conforms to the social responsibility model; rather it is a mixture of the two (Caspi & Limor, 1999). In its early days, the Israel Broadcasting Authority (IBA) adopted the state-owned public media model of the BBC, while the press was an organ of political parties, a workers cooperative, or a family business. During its first 30 years, Israeli media followed the state's political leadership, supporting the struggle for independence amidst neighboring hostile countries and accepting, willingly, tight censorship for national security's sake. From a functional point of view, the media institution was founded at the same time as other state institutions and "grew up" with them, taking an active part in shaping the political reality (ibid.).

Israeli mass media also identified with the economic policies of the political leadership. With origins in East Europe, the immigrant-leaders who founded and developed the society that became the state of Israel applied a socialist ideology. Americanization and globalization trends infiltrated Israeli society gradually. Indeed, it was only in the 1990s that commercial television channels began broadcasting and soon after portions of the print press were purchased by international media corporations. While the IBA continues to broadcast alongside commercial media, it is deeply politicized and much less popular with audiences (ibid.).

In 2002, state intervention stopped media corporations' monopoly of media channels. They did so by tendering broadcast licenses to parties linked to national business corporations. The fears of opponents of this policy have been realized as corporate interests have interfered with broadcast content when it did not suit their commercial interests (Drub-Heishtein, 2006). At present, competition between media organizations in Israel is fierce and violent. As a result, power struggles often take precedence over the public's right to know (Kanti, 2007).

These circumstances may explain why the national mass media ignore political bloggers as a factor in the Israeli media ecology. Media corporations interpret blogging solely as an informal style of personal columns, as is the case of "blogs" contributed by leading journalists on their newspaper's Web sites. Indeed, the few political journalists who blog independently either recycle their printed stories or choose to blog about any topic other than their professional niche. Ironically, the j-bloggers who often comment on political issues are technology or culture media professionals.

The national mass media was forced to relate to an anonymous j-blogger nick-named Velvet Underground in 2006 who published institutional gossip and behind-the-scenes stories about the local media industry. Velvet Underground turned out to be the vice editor of an entertainment magazine famous for her cinema critiques. She continues blogging under her real name, reaching beyond her professional niche and often addresses key political issues. Here, too, there is almost no mass media follow-ups on her political stories.

On a few occasions, the mass media reprinted stories that appeared on Velvet's blog or on *Peer Pressure* without giving them proper, professional credit. In an online public debate about this matter, one journalist spoke in defense of his media organization in explaining its policy against crediting blogs: "A blog isn't a professional source and a journalist who blogs isn't acting within his profession at the time. It is as ridiculous as crediting rumors and private conversations." This view suggests a third explanation why political blogs are ignored or, at best, mentioned as social initiatives: Blogs are perceived as an expression of the writer's identity as if it is a form of interpersonal interaction within a virtual community representing the public sphere. In this view, a blog is no different than a forum. As such, blogs serve as the contemporary version of the nineteenth-century give-and-take exchanges that took place in European cafés and through which public opinion was formed and reported by journalists (Tarde, [1898]1969). Such an interpretation neutralizes and depoliticizes blogger authorship.

As a result of this situation, political blogging in Israel is normally considered to be a facet of Internet culture and as such covered solely by computer and Internet media channels. Indeed, *Peer Pressure* was the only political blog to receive any media coverage outside computer and Internet media channels. However, the coverage was depoliticized. For instance, though YNet[5] rewarded *Peer Pressure's* in-depth coverage that led to an organized blogger protest against the Omnibus spending law[6] as "one of the best social initiatives of the year," it placed the story in the less popular social involvement channel.

A fourth and final explanation for ignoring political blogs has to do with the political orientation of Hebrew-language bloggers. Although political blogs in the United States are almost equally divided between liberal and conservative points of view (Adamic & Glance, 2005), blogs responsible for the key journalistic success stories are conservative ones such as *Andrew Sullivan* or *Powerline* that target left-wing media organizations and support the Republican administration.

In Israel it is interesting to note that given the profound institutional changes within Israeli media organizations over the last decade, they seem to take a right-wing stand on economic policies, backing recent trends of privatization led by the current administration. In contrast, the vast majority of Israeli bloggers who comment on political issues promote a left-wing socialist agenda regarding security as well as social and economic matters. This finding was empirically backed during the 2006 elections when bloggers surveyed the blogosphere and in one post reported how politically identified bloggers voted. This strong tendency may explain the fact that political bloggers' initiatives are neither encouraged nor extensively covered by mainstream media in Israel.

Blogging as an Ideal Type of Journalism

Anecdotal evidence gathered reveals that blogging is often portrayed as an ideal type of journalism in regard to journalistic ethics. In a debate over blog-branded content on *Peer Pressure*, authors and many other commentators stressed the importance of blogger independence, stating that they embrace the ideal journalism ethic that in their view has disappeared in media organizations. However, several other j-bloggers argued that only the involvement of advertisers will enable the blog genre to have substantial influence and to attract more audiences.

Here, there seems to be agreement between political bloggers and media researchers in Israel that entertainment has overshadowed all other media roles, turning commercial interests into professional norms. Bloggers contend that this transformation has left the more serious, professional journalistic endeavors to bloggers who are free of special interests. Though the blogs examined in this study are considered important and valued by both bloggers and mass media practitioners covering online activity, they are different from American political blogs; they suffer from relatively low readership and rare media coverage. None of these blogs is involved in gathering news stories or checking facts, rather they engage in correlation and mobilization practices generally associated with media roles in the functional theory media literature.

Peer Pressure is written by Yair Tarchitsky and Yoav Ribak, two editors of *Haaretz,* a daily newspaper. The authors promote a left-wing socialist agenda through cumulative counter readings of news and events as well as exposés on faults of the social system that provide, according to the authors, more complete analyses than the fragmented interpretations presented by the mass media. Their most noted activity was leading an organized blogger campaign to cancel the Israeli Omnibus Spending Law. *Peer Pressure* is one of the political j-blogs that resemble the American model. However, this j-blog, like other personal journals kept by Israeli journalists differs from its American counterparts in that it is engaged in correlation and mobilization rather than solely media surveillance. The following is the opening statement of the *YNet Social Gaps* forum written by the authors of *Peer Pressure,* who also manage the forum:

> As journalists, we feel the mass media does not handle properly issues that fall between politics and economics. Fair distribution of resources and workers' rights have become bad words and social issues are reported superficially or from the wrong perspective. We decided to create a blog where we can write freely about the issues we find important and believe should be on the public agenda [. . . [

Examination of the blog's contents reveals that *Peer Pressure* authors do not compete with mass media over news stories, rather their activities seem to advance what they consider to be good journalism. Such activity is less about information gathering and more about "correlation"; that is, the way information is presented and interpreted for dual purposes: to educate and to mobilize. Involving personal stories of deprived workers or segments of society neglected by the government, their original stories may seem important but are seldom considered to be newsworthy according to newly transformed mass media norms referred to previously. In an interview, Yair Tarchitsky of *Peer Pressure* noted:

> The media do not change things. They report events and interpret them superficially. We do not present news in the blog. We would rather explain things people already know. The blog has a different rhythm. It is free of time limitations or frame manipulations, so we can bring in-depth stories, continuously. Nobody can say these are not worth publishing, since a similar story already exists. [. . .] On a couple of occasions we stumbled across exclusive pieces of information but we decided to leak them to the media. Gossip and scandals only draw attention away from our goal.

At the heart of the arguments and actions by *Peer Pressure* authors is their concern about the profound commercialization of the media and their hegemonic practices, disguised as professional journalistic norms. Accordingly, j-bloggers are not trying to change media organizations, rather they are involved in creation of an independent alternative medium. Thus, while "breaking news" has taken on a meaning of breaking the routine, the objective of *Peer Pressure*'s correlation practices (in terms of the role of the media) is to deconstruct "routine." For example, their online campaign to cancel the Omnibus Spending Law stemmed from a series of correlative postings tying sporadic events into one context and deconstructing the meaning of this extremely complex, seemingly mundane law into concrete daily actions that affect the entirety of society. In doing so, they redefined "routine" as a serious problem. While initially the intent was to be educative, the correlative approach evolved into spontaneous mobilization. This process could have been framed as its own news story.

An additional example is the blog *KnessetWatch*, whose purpose is to engage in participatory legislative critique. Hoping to advance a vibrant public debate, it summarizes and explains the meaning and context of new legislative proposals whose source is the official Knesset Web site. The author, Dubi Kanengisser, a graduate student in political science and Web veteran, invites all readers to participate in the game of legislative criticism and occasionally emails the link to relevant politicians. In doing so, *KnessetWatch* offers a practice important in a democracy that should be performed by mainstream mass media, in their role as "watchdog" of the legislative process. In a posting on his personal blog, Kanengisser wrote the following about *KnessetWatch*:

> I think the principle of *KnessetWatch* is important and belongs in the mainstream mass media [where it will attract] more exposure than my 50 entries [. . .] Newspapers should have legislation critics, though they might receive instructions to be cynical and mock the proposals, and if so this would miss the point. But at least there would be a public debate over legislation, exactly like there is over a stupid new TV show [. . .]

In terms of blogs' media roles, legislative critiques and the public debate about legislation could also become correlation practices if they result in latent mobilization. This would happen if explanation and discussion of the implications of a legislative proposal lead to initiation of direct contact between bloggers and politicians. Kanengisser's statement reveals a concern for the traditional roles of the mass media being overshadowed by the role of entertainment. In response, his blog establishes a genre that in the former view of journalism ideally belongs in a newspaper. Interestingly, in our interview, he admitted that he hopes the blog will evolve into a mainstream newspaper column or at least inspire such a column.

Inspired both by Landa's email initiative and the American Web site *Democratic Underground*, *Black Labor* is an online attempt to organize Labor Party voters in order to establish direct email contact with Party representatives in the Knesset, followed by publication of such correspondence on the blog. The origins of the blog are in Labor Party voters' anger and disappointment immediately after the 2006 elections when the Party's leader, elected on his record for social activism, chose to become Defense Minister, an act interpreted as abandonment of this agenda. Led by activist Itay Asher and blogger Yohai Eilam, 30 high-profile bloggers joined the initiative, including a number of journalists and the nephew of the Labor Party leader. Blog readers were invited to propose pragmatic topics and email their proposed letters, regardless of political affiliation.

Since the blog usually calls for action from people who are already able to define the problem, it seems that *Black Labor*'s main interest is mobilization. By combining features of mass media and activism, it demonstrates how a blog can serve simultaneously as a publication and as a congregation; that is, as a site for meeting in a community space. Indeed, many of the *Black Labor* group are j-bloggers who have their own separate personal blogs in which they often write political and social critique. However, the founder and leader of the blog group, Itay Asher, regards blogs mainly as a mobilization tool and means of congregating. In the interview, he challenged the influence attributed to commentaries and other correlation practices:

People blog their opinions on events. Everyone wants to be a Joel Marcus or Doron Rosenblum [columnists for *Haaretz*], but most don't have the talent, nor are they read by people other than a small community of bloggers. [. . .] Even if it's a good interpretation it has no influence. Also, I can't quantify the influence of an article by Marcus [. . .] the key to having an influence via the web is networking, organizing, connecting people.

Asher represents a pragmatic action-oriented view of blogs as gathering spaces with an inherent communality that is potentially political by nature, using networking as a synonym for unionizing. Though links to a blog are normally interpreted as recommendations or a sign of friendship, Asher refers to bloggers linked with *Black Labor* as "our supporters." However, by offering a critical model for participatory journalism, *Black Labor*'s blogging style blurs direct contact with politicians with journalistic professional norms, as stated by Asher:

Politicians have no real information on how we expect them to act on our behalf. The media pretend to know our will, but they don't. We must communicate directly. We need to watch them personally between elections so they keep to their campaign promises. [. . .] Political discourse in the media is mere gossip. For instance, I asked Ami Ayalon to state five things he intends to do as Defense Minister, while journalists asked him about his strained relationship with the current minister or about his childhood. It's like a soap opera; there are no reports on practical actions.

Asher feels the media have failed in their most fundamental role as watchdog of democracy, trading professional journalism norms for entertainment and scandals that sell newspapers. Therefore, he thinks citizens should demand accountability from politicians in an alternative way. His critique also addresses the notion of "good news," claiming the media do not cover all aspects of life in a creative manner, only public affairs that go wrong. *Black Labor* sees itself as an alternative, fulfilling the gatekeeper role of participatory journalism. However, blogs' gatekeeping and mediation practices reach beyond normative media discourse. In the case of *Black Labor*, blogs can act as surrogate representatives of Labor party online institutions. For example, since the Labor Party Web site is rarely updated and lacks basic information for potential new members, *Black Labor* has published such data several times and interacted with citizens requesting information about the Party online. Another result of blurring political activism with media roles, this particular form of gatekeeping, too, confuses some bloggers as to the purpose of this blog.

Interestingly, soon after completion of the field portion of this research, *Black Labor* relocated to a Wordpress-based template. This suggests that it has outgrown its grassroots nature and has declared that it wants to function as a Web site, although it continues to share blog characteristics. Furthermore, a survey circulated sought to ascertain group members' views of *Black Labor*'s purposes as a community of activists or as a media center. Further research should follow the evolving crystallization of this unique blend of grassroots activism and participatory journalism in Israel.

Assessing Israeli Blogosphere's Influence

The research of blogs cited previously measured American blogs' impact on national mass media practices and on the national political agenda. Just as these studies were conducted largely through anecdotal evidence and traffic indicators, the findings presented applied the same indicators to assess the influence of Israeli political blogs, while pointing to possible opportunities for long-term impact.

Measuring influence within the blogosphere has used comments, trackbacks, subscribers, and links from other blogs as blog traffic indicators. This approach has been compared to measuring academic influence by counting citations within a field of study (Gill, 2004). *Technorati* has been found to be the most comprehensive tool available for measuring the number of links to a blog (ibid.), however it does not cover the major Israeli blog-hosting Web sites. This obstacle was resolved by blog authors sharing their administrative tool links. In addition, since not all links are created equal, influence cannot be considered to be solely a matter of traffic. Therefore, complementary measurements of influence were applied including mass media coverage and politicians' responsiveness.

Table 7.1. Influence indicators of the blogs examined*

Blog	Months online	Posts	Comments	Trackbacks & Permalinks	Total visitors	Media	Subscribers**
Knesset Watch	8	46	209	35	3,062	1	8
Peer Pressure	11	108	1,758	367	34,343	7	191
Black Labor	9	374	1,801	379	22,379	4	30

*From date of creation until end of March 2007.
**Does not include many RSS subscribers who cannot be accounted for.

As seen in Table 7.1, *Peer Pressure* attracted the highest blog traffic although its authors were much less active than *Black Labor*. *Peer Pressure* is also the only Hebrew political blog that attracts readers and commentators outside the blogosphere. This may be due to the mass media coverage it has received. *Knesset Watch* is noted for low activity indicators. In this case, there seems to be a correlation between the activity of the author and the readers, with the former being discouraged by the readers' low traffic. Kanengisser admitted in the interview that only two other bloggers emailed him suggesting legislation to be reviewed, but none came through when asked to author the posts themselves.

Knesset Watch appears on the relatively new hosting Web site *Blogli,* while *Peer Pressure* and *Black Labor* appear on the popular *Isra-blog* Web site. The communal aspect of the *Isra-blog* host Web site may explain the many trackbacks that represent debates within the blogosphere, while *Knesset Watch* has failed to generate such a debate record. I found at least two possible opportunities for cooperation between *Knesset Watch* and *Black Labor* that were missed although the blogs are linked to one another, for example, by directing a legislation critique into an inquiry email.

Black Labor and *Peer Pressure* members have cooperated in crafting their blogging activities: Tarchitsky of *Peer Pressure* used his personal contact with Labor Party members and their assistants to draw their attention to *Black Labor,* while *Black Labor* bloggers actively promoted the anti-Omnibus Spending Law campaign. The bloggers occasionally coordinated action by telephone, and some symbolic action materialized when one *Black Labor* blogger took anti-Omnibus Spending Law posters to a national social action convention attended by party members and activists. Though linked and supported by fellow bloggers, Kanengisser of *Knesset Watch* did not actively tap into this backstage blogging network.

American success stories prove the importance of links and trackbacks as indicators for networking and cooperation between blogs in impact-building and sustaining a story. Almost all blog-originated stories made their impact only after being linked to and debated on several leading blogs. However, overall, all three political Hebrew blogs suffer from relatively low readership and traffic. This suggests that new engagement opportunities online will not necessarily attract people who have little interest in politics (Davis, 1999).

In 1999, Davis (ibid.) claimed that politicians show low rates of online responsiveness. This is a situation that had the potential to improve with introduction of blogging. *Black Labor* had high responsiveness from politicians and showed that it had the ability to affect the political agenda in the short term. *Peer Pressure* attracted more media coverage and may have had a possible impact on professional peers and the public opinion, suggesting that it may have the capability to induce change over the long run. In general, it is hard to determine which of these influence models is more significant.

Fourteen out of the nineteen targeted Labor Party Knesset members regularly answer *Black Labor* emails and address their concerns. Most have their assistants follow the blog on a regular basis. However, aside from having the ability to create a "watchdog" environment, there were no unmitigated success stories. So far, only one concrete parliamentary success has been linked to the *Black Labor* group: Blogged personal testimonies of Israeli Post Authority contract workers led to three mass media items on their poor working conditions, one of which was a prime-time televised report. While a special parliamentary debate on the matter was called, it is difficult to know if it was the mass media coverage or emails sent to Labor

Party Knesset member Shelly Yechimovich by *Black Labor* bloggers that were responsible. The mass media coverage ignored *Black Labor's* role in the story, and some bloggers complained the story was constructed as a single news event, ignoring the habitual national problem of the working conditions of all contract workers. Yet, they admitted that this was a good start as the first and so far the only political success story that started on a blog in Israel.

Though *Peer Pressure's* campaign against the Omnibus Spending Law attracted more media coverage and blog links, only slight changes were introduced to the law. Tarchitsky stated in the interview that he has no idea if the Knesset member advancing the changes even heard of the campaign, since the law was severely criticized by other Knesset members. However, *Peer Pressure* explained the meaning of this seemingly routine law to each and every citizen in a way un-precedented by the media and helped raise public awareness about similar routines, at least among its readers: Authors received emails from readers, including fellow journalists, in appreciation for their assistance in improving their understanding of the actual consequences of developments and events. The authors testified that their status as journalists rose in the eyes of their peers. Further, they felt that being read by peers might affect media coverage in the long run, an effect that they value more than direct coverage of their blog topics. In contrast, *KnessetWatch* attracted neither media nor politician attention. Kanengisser related that though he had emailed a link to a Knesset member occasionally, he had not yet received an answer.

The influence indicators reviewed correlate with characteristics of the blogger's identity, including perceived status and imagined representation. Knesset members responded to j-bloggers, party members, or a perceived organized group. They also responded to angry or emotional emails concerning burning issues, while ignoring the formal logical claims concerning legislative routines of a non-networked blogger.

Close Encounters of the Unmediated Kind: Politicians as Bloggers

The increasing significance of the American blogosphere noted may also be due to the growing number of American politicians becoming bloggers or interacting with bloggers online, especially prior to elections. The following is an account of Knesset members' attempts at blogging and summarizes their view of the roles of blogging and the political significance of the Israeli blogosphere.

In general, Israeli politicians have little awareness of Internet technologies as an infrastructure for mass media. Exceptional in this regard is Likud party member Michael Eitan who was the first politician to promote information technology policies. Indeed, long before the term *Web log* was coined, he managed a frequently updated professional Web site that is still active. Ehud Yatom of the Likud Party (moderate right-wing) was the first Knesset member to attempt blogging. In reply to one of Landa's queries in 2003, Yatom explained that he did not report some of his activities to the public because of a lack of mass media interest. A second blogger who commented on this blog-email correspondence suggested to Yatom that he also blog in order to address the public directly about such issues. And, indeed, Yatom did ask bloggers to help his assistant create such a blog. On February 23, 2004 the blog was created by Yatom's assistant using the popular *Isra-blog* software.

The blog's heading stated its purpose as "expressing Yatom's opinions without mass media mediation." Direct dialogue with the public, however, proved dysfunctional: Bloggers ignored the content of the first post and chose to express their harsh and even rude criticism of Yatom over a national security scandal he had been involved in a number of years before. Yatom's assistant closed the blog immediately. The affair was picked up by all computer and Internet media channels. Yatom told the press he hadn't backed off because of the comments, but because he hadn't authorized the blog to begin with. His assistant resigned following the incident.

A few bloggers called on Yatom to reopen his blog, including one blogger who had posted harsh comments on the blog. He cited the email he sent to Yatom in his personal blog: "I admit my comment was irrelevant and ugly, however this affair would have faded rapidly and you would have had the opportunity to be the first politician to blog and strengthen your relationship with both your voters and your opponents" (transcripts of *Boles Shikmim* deleted blog).

Israeli political discourse, as reflected in the media, is notoriously fierce and overtly confrontational. Research has offered evidence of its resemblance to the oral tradition of Talmudic text argumentation that uses the adversarial format to maximize mutual comprehension and ultimately enhance sociability (Blondheim, Blum-Kulka, & Hacohen, 2002). The response cited above exposes the blogger's latent assumption that sociability would ultimately be enhanced despite his rude attack, stressing the importance of the mere existence of a direct dialogic channel with politicians over its ephemeral confrontational content.

In 2005, Shaul Yahalom of the Mafdal Party (Orthodox right-wing) was invited to blog by the *Kipa* religious community portal. Yahalom's choice of blog location resulted mainly in closed discourse within the religious community. He did not respond to the comments received, either directly or in his next post. He stopped blogging when he failed to be re-elected to the Knesset.

During 2005, Roman Bronfman of Democratic Choice and Dov Chanin of Chadash, both extreme left-wing parties, created Wordpress-based blogs. Both politicians were working with new media consultant Guy West, one of the founders of *Indymedia*. Bronfman's blog created a small active community but ceased to update when he decided to withdraw from the elections. Chanin's blog is still online; however it is rarely updated and has not managed to engage in a vibrant dialogue, though many bloggers list it on their blog roll.

Table 7. 2. Traffic on Knesset members' blogs

Politician	Political orientation	Months online	Posts	Comments	Blog links & trackbacks*
Shaul Yahalom	Religious right-wing	12	25	349	2
Roman Bronfman & team	Extreme left-wing	6	41	295	67
Dov Chanin & friends	Extreme left-wing	11	176	202	130
Arie Eldad	Extreme right-wing	2	10	251	4
Shelly Yechimowich	Moderate left-wing	2	9	No option	1

*Shows only independent blogs due to Technorati limitations.

Since both politicians have very similar political views and their blogs were produced, updated, and managed similarly by the same advisor, the only variable that explains the deviation between the two as seen in Table 7.2 is their personal writing style and responsiveness to their readers. Bronfman himself was highly involved and interested in the blog and his staff responded to comments on regular basis; Chanin did so only in the very beginning. A large portion of the posts on Chanin's blog are recycled mass media articles and many of the posts contain no comments.

On October 12, 2005, Bronfman posted: "I fired my publicity agent. Anyone who blogs for the Democratic Choice Party should blog for himself. The blog is a literary genre like a journal. I told my spokesperson that blogs should be written in the first person."

A handful of politicians attempted to use blogs as a propaganda tool before the elections of February 2006. Most blogs appeared within party Web sites, had no options for comments, and were updated by writers who misinterpreted the informal blogging style for a gossip column genre and were therefore ignored by most bloggers. Only Shelly Yachimovich of the Labor Party, an ex-journalist, was "excused" by the bloggers since she updated her page personally.

Bloggers' conversations about these blogs made it clear that they have no reason to visit a blog that is merely a cosmetic change of format for propaganda.

Bloggers expect the blog to either provide direct communication or, at the very least, be updated directly by the politician. Without either of these proximity indicators, a blog has no added value over a personal column, a paid ad in the mass media, or an informational Web site, even if it is technically referred to as a blog.

Owners of the *Isra-blog* Web site invited politicians to blog during the 2006 election campaign. The only current Knesset member to create a blog was Professor Arie Eldad of the Ichud Leumi Party (extreme right-wing). Eldad had a genuine talent for writing and the patience to respond to most commentators. He seemed to enjoy every minute of blog engagement and exploited all the possibilities of the medium. Therefore he was embraced by bloggers, despite his perceived extreme opinions. As seen in Table 7.2, his blog shows high traffic during the short time he blogged.

Content and blogosphere connectivity were enhanced when Israeli politicians allowed comments and responded to them in their blogs, and they also gained an indicator of proximity. In contrast, only 43 percent of American blogs included comments when first sampled; consequently, comments were not a defining feature of blogs (Herring, Scheidt, et al. 2004). American bloggers think that while comments are not a requirement, they do enhance content dramatically (Arrington, 2006).[7]

Several bloggers expressed relief from anxiety due to their sense of proximity to the politicians as a result of Eldad and Yahalom's personal blogging and/or responsiveness. This is noteworthy since these politicians are perceived as having extremist political views. However, overall, all Israeli politicians' blogs show low traffic indicators, lower than even non-politician political blogs. This is a trend that undermines the blogosphere's chances to serve as a political arena in the next elections.

By April 2006, two months after the last elections, none of the above politicians was updating his or her blog. In March 2007, shortly after the field research was completed, two leaders of mainstream parties started routine blogging: Likud Party leader and former Prime Minister Benjamin Netanyahu and leader of the left-wing Meretz Party, Yossi Beilin. While Netanyahu blogs in a carefully edited speech-style of rhetoric on a private domain and ignores readers' comments, Beilin blogs in a personal journal style within the new *The Marker* social network and often responds to comments. Further research should consider comparing Netanyahu and Beilin's distinctive blogging practices as they unfold, to determine whether the role of personal interaction is indeed crucial for a blog's political value, from the politician's perspective.

Being Good Neighbors: The Blogosphere as a Dialogic Space

Given the focus on economic and social issues, the majority of routine political writing in Hebrew concerns internal politics. I maintain, however, that the discursive term "the Israeli blogosphere" consists not only of Hebrew-language blogs but also of English-language blogs written by Israelis. In fact, these blogs represent the Israeli blogosphere for non-Hebrew readers.

Before mid-2006, these blogs were difficult to track since they appeared on various international platforms. During the war in Lebanon in the summer of 2006, a voluntary aggregation index of Israelis blogging in English was created and continues to be updated.[8] Linking and aggregation is a grassroots gatekeeping mechanism creating discursive fields that result in inclusion and status conferral. Thus, while the index currently lists feeds from 169 blogs, these do not seem to represent the majority of Israeli bloggers in the English language as the vast majority of these blogs present a left-wing agenda. Indeed, a few bloggers I've spoken to have concluded that they are a minority among the right-wing extremists blogging from settlements in the Occupied Territories, to which the voluntary aggregator refuses to link.

These 169 blogs enjoy participation in blogosphere discourse both on the Israeli and the international level. For example, 2,666 international blogs in various languages link to the most popular English-language Israeli blog, belonging to j-blogger Lisa Goldman.[9] Although English-language Israeli blogs were not included in the corpus of this research, a review of the Israeli blogosphere would not be complete without a brief overview of their activity:

> English-language, Israeli blogs address foreign policy and Palestinian authority-related issues much more than they do internal economic or social issues.
>
> Whereas the most valued political bloggers in Hebrew are male, some of the most influential English-language bloggers are female.
>
> Most bloggers perceive the purpose of their blog as a window into everyday life in Israel.

It is through this extreme subjectivity that these blogs exercise a form of communion with possible political implications. While Hebrew-language political posts circulate within a small circle of readers, the daily experiences and conversations of English-language Israeli bloggers attract much more attention, open dialogues, shape opinions, and offer alternative narratives to Israel's mass media image. Their political importance does not lie in the content of their posts, but in the mere fact they are blogging in the context of the international blogosphere. Israel is engaged in a historical dispute with its Middle East neighbors. Even when there are formal peace treaties between Israel and some of its neighboring countries, the majority of the populations in these societies is misinformed due to biased coverage in the national mass media and may be hateful to Israelis both offline and online. Yet, one could argue that the Internet presents an unmediated opportunity for the sides to meet and discuss. Blogging contributes to this opportunity by allowing people to share continuously from the depths of their life experiences and personality.

Further, since blogs are personal spaces, people from both sides literally visit each other's forbidden territory. The Israeli-Lebanese war of 2006 inspired some Israelis and Lebanese to author blogs together. Beyond symbolizing and merely engaging in an online dialogue, this was an act of mutual creation, shared space, authorship, and responsibility. As a shared space, such a blog becomes binding; "it is like moving in together," as one Israeli blogger noted in an informal conversation. Anecdotal evidence shows that bloggers were able to translate this metaphor into friendships that reshaped opinions. Such friendships between j-bloggers and political activists in Middle East countries may turn out to be valuable networking resources for their countries.

The metaphor of space has become an actual geographic space on several occasions. For example, over the last two years several bloggers from countries such as Iran and Lebanon have visited their blogger friends in Israel using foreign passports. While most of these visits have been conducted in a discrete manner, at least two such visits were covered extensively by mainstream mass media, including a televised item. This demonstrates that blogs not only expand mass media coverage and break its framing but also invite us to engage in transformative personal interaction beyond being merely informed.

The latest initiative in which I am an active participant-observer is *Good Neighbors*,[10] a group blog coauthored by leading bloggers from various Middle East countries who are attempting to reside in and create a community in mutual space. This construct alone is politically valuable regardless of the contents of the blog.

Discussion and Conclusions

To date, research that has focused on changes in the mass media attribute their decreasing credibility to political and economic constraints (Baldasty, 1992; Mosco, 1996), conformity to hegemonic ideologies (Bennett, Gressett, & Haltom, 1985) or abandonment of public discourse to superficial entertainment (Postman, 1985). While hopes were expressed that the net would offer alternative media models, initial research suggests that offline media reproduced themselves online (Mansell, 2004; McChesney, 1998). Bloggers may be in a position to undermine this process, acting as independent volunteers free of production costs, interests, or popularity concerns.

The ideology of public journalism sees its task as not only informing citizens, but also enhancing meaningful public discussion and participation (Rosen, 1991). In the United States, what seemed to be the emergence of a new postmodern practice of journalism is gradually being abandoned by j-bloggers as they conform to mainstream journalism norms (Robinson, 2006). However, in Israel, leading political blogs seem to be seeking to attain an ideal model of serious, truly independent journalism. In doing so, they retrieve traditional facets ideally performed by media practitioners and re-establish pre-commercialization genres in new media contexts. Such efforts are historic, at least in the local context. Thus, while American bloggers make news and uncover scandals, Israeli political bloggers attack political routines through correlation and occasionally mobilization. In doing so, Israeli political bloggers try to take up a task neglected by the mass media.

This practice resonates in part with the role of public property media. Like public property media, Hebrew political bloggers voluntarily represent worthy causes and practices that should exist in the mass media, regardless of popularity concerns. Ironically, the Internet and its blogs are public property media in the most direct sense. Therefore, we should expect blogging to perform everything people perceive as necessary or worthy but cannot realize due to institutional, commercial, and other constraints. Such potential alone points to the political significance of blogging as a practice, regardless of blog content.

Furthermore, blogs are blurring the boundaries between publishing media and public space. Therefore, bloggers most realize the potential of blogging when their practice moves between the metaphors of media and political participation; for example, correlation overlaps mobilization when posts generate and engage in networked dialogues. Most Israeli politicians, however, seem confused by these overlapping dimensions. Their blogging is propagandistic, as if it were another mass medium, ignoring its dialogic aspects. However, interestingly, some have related to dialogic attempts of other bloggers as voices of citizens rather than as fellow authors or semi-journalists.

The short experience of Israeli politicians with the Hebrew blogosphere indicates too small a readership on a national scale to justify an investment in personal blogging. This kind of personal, informal interaction seems to attract and perhaps benefit politicians with extreme views who seek to soften their image and to relate to the public in a friendly manner. However, blogging seems of less value and may even prove to be a dysfunctional burden for a mainstream politician, maintaining the importance of gatekeeping mechanisms versus the Israeli fierce political discourse and conversation norms.

Alternatively, interacting with organized bloggers, in a group setting such as *Black Labor*, could provide such a mechanism for politicians who want to interact directly with the online public without making a commitment to invest in establishing and maintaining a personal blog. Bloggers should be reminded, however, that being independent doesn't equal being neutral. At least in the case of *Black Labor*, the blurred nature of the blog might result in a clash between the watchdog role and the gatekeeping model which is similar to the historical party-owned newspaper. In this case we see that the bloggers' group is often torn between "barking" at party members and representing the party online.

While American mass media might pick up a single blog story and enhance it, the absence of attention or credit from Israeli mass media sheds light on the blurred features of mass media and interpersonal communication within blogs. The importance of community networking and personal dialogue in their online form suggests that the key to understanding the blog's role is not necessarily within mass media frameworks and may even be anecdotal to them. Thus, while a single mass media item might have an immediate impact, it seems the blogosphere as a media space involves cooperation between many Internet sites that link, literally, the *Web* of a story.

In regard to mass media practices, Israeli blogs are unable to compete with mainstream journalists and do not attempt to do so. Indeed, the blurring of the boundary between practices of mass media and political participation suggests a model that may enable the Israeli blogosphere to become an independent, possibly influential player, without requiring the help of the mass media. The Hebrew blogosphere seems to be evolving slowly and may still resemble its American counterpart. However, readership and the perceived status of blogging must increase dramatically before this may happen. Further ethnographic research on the blogosphere, I believe, may indicate that the modes of action and experience blogs facilitate, rather than the content of blogs, are key indicators in measuring their political influence.

Notes

1. All interviews and Hebrew-language blogs were translated by the author.
2. Blogs not updated for over two months were considered inactive. Data were received from blog-hosting Web site owners, while independent blogs were traced through *Technorati*. Active blogs form 20 percent of the overall quantity of existing blogs (according to data supplied by *Isra-blog* and *Tapuz's blogosphere* Web sites owners).
3. http://www.wordpress.com
4. In terms of the Israeli political context, the right-wing is conservative in terms of security matters but liberal in terms of social and economic affairs, while the left-wing seeks security amelioration and has a tradition of socialist views of social-economic issues.

5. The most popular Israeli online newspaper run by the popular *Yedioth Ahronoth* daily newspaper.
6. The Omnibus Spending Law is a unique legislative arrangement allowing economic officials the power to conduct vast reforms such as budget cuts and privatization. The law was necessary in 1985 to stabilize an economic crisis; however, its critics claim it is unnecessary and even harmful at this point in time.
7. The writer thought blogs are conversation and suggested that a blog without a *comments* option is not a blog. However, he added a poll to his post on *Techcrunch*, and bloggers voted otherwise.
8. http://english.webster.co.il
9. http://ontheface.blogware.com
10. http://gnblog.com

Bibliography

Adamic, L. A., & Glance, N. (2005). The political blogosphere and the 2004 U.S. election: Divided they blog. Workshop on the Webloging Ecosystem, WWW2005. Retrieved December 19, 2006 from http://www.blogpulse.com/papers/2005/AdamicGlanceBlogWWW.pdf

Arrington, M. (2006, December 31) *Techcrunch: What is the definition of a blog?* Retrieved December 31, 2006 from http://www.techcrunch.com/2006/12/31/what-is-the-definition-of-a-blog/

Baldasty, G. J. (1992). *The commercialization of news in the nineteenth century.* Madison, WI: University of Wisconsin Press.

Bennett, W. L., Gressett, L. A., & Haltom, W. (1985). Repairing the news: A case study of the news paradigm. *Journal of Communication, 35*(2), 50–68.

Blondheim, M., Blum-Kulka, S., & Hacohen, G. (2002). Traditions of dispute: From negotiations of Talmudic texts to the arena of political discourse in the media. *Journal of Pragmatics, 34*, 1569–1594.

Caroll, B. (2004). Culture clash: Journalism and the communal ethos of the blogosphere. In L. Gurak, S. Antonijevic, L. Johnson, C. Ratliff, & J. Reyman (Eds.). *Into the blogosphere: Rhetoric, community, and culture of weblogs* [Electronic version]. Retrieved 16 March, 2005 from http://blog.lib.umn.edu/blogosphere/

Caspi, D., & Limor, Y. (1999). *The in/outsiders: The media in Israel.* Cresskill, NJ: Hampton Press.

Davis, R. (1999). *The web of politics: The internet impact on the American political system.* New York: Oxford University Press.

Drub-Heishtein, G. (2006). Was the struggle against cross-ownership a mistake? [in Hebrew]. *The Seventh Eye, 61*(3), 55–58.

Gallo, J. (2004). Weblog journalism: Between infiltration and integration. In L. Gurak, S. Antonijevic, L. Johnson, C. Ratliff, & J. Reyman (Eds.), *Into the blogosphere: Rhetoric, community, and culture of weblogs* Retrieved 16 March 2005 from http://blog.lib.umn.edu/blogosphere/weblog_journalism.html

Gill, K. E. (2004). *How can we measure the influence of the blogosphere?* WWW2004 Convention, New York, May 17–22, 2004. Retrieved September 6, 2006 from <http://faculty.washington.edu/kegill/pub/www2004_blogosphere_gill.pdf

Herring, S. C., Kouper, I., Scheidt, L. A., & Wright, E. (2004). Women and children last: The discursive construction of weblogs. In L. Gurak, S. Antonijevic, L. Johnson, C. Ratliff, & J. Reyman (Eds.), *Into the blogosphere: Rhetoric, community, and culture of weblogs.* Retrieved 16 March 2005 from http://blog.lib.umn.edu/blogosphere/women_and_children.html

Herring, S. C., Scheidt, L. A., Bonus, S., & Wright, E. (2004). *Bridging the gap: A genre analysis of weblogs.* Proceedings of the 37th Hawaii International Conference on System Sciences (HICSS-37), Los Alamitos, IEEE Computer Society Press.Retrieved January 3, 2006 from http://www.blogninja.com/DDGDD04.doc

Johnson, E. (2006, November). Democracy defended: Polibloggers and the political press in America [Electronic version]. *Reconstruction: Studies in Contemporary Culture, 6*(4). Retrieved 2 December 2006 from http://reconstruction.eserver.org/064/johnson.shtml

Kanti, N. (2007). Competition without borders: Channel 2 against channel 10, Yedioth against Ma'ariv [In Hebrew]. *The Seventh, Eye 66*(1), 20–23.

Koren-Diner, R. (2006, November 28). TIM survey: Internet users in Israel reached 3.9 million [Electronic version in Hebrew]. *Haaretz [Tel Aviv].* Retrieved November 28, 2006 from http://www.haaretz.co.il/hasite/pages/ShArt.jhtml?more=1&itemNo=793767

Lowrey, W. (2006). Mapping the journalism-blogging relationship. *Journalism, 7*(4), 477–500.

Mansell, R. (2004). Political economy, power and new media. *New Media Society, 6*(1), 96–105.

McChesney, R. W. (1998). The political economy of global communication. In R. W. McChesney, E. M. Wood, & J. B. Foster (Eds.), *Capitalism and the information age: The political economy of the global communication revolution* (pp. 1–26). New York: Monthly Review Press.

Mosco, V. (1996). *The political economy of communication.* London: Sage.

Norris, P. (2001). *Democratic phoenix: Reinventing political activism* [Electronic version]. Cambridge: Cambridge University Press. Retrieved January 2, 2007 from http://ksghome.harvard.edu/~pnorris/Books/Democratic%20Phoenix.htm

Pew Internet & American Life Project (2006, July 19). Retrieved November 23, 2006 from http://pewresearch.org/pubs/236/a-blogger-portrait

Postman, N. (1985). *Amusing ourselves to death: Public discourse in the age of show business.* New York: Viking.

Robinson, S. (2006). The mission of the j-blog: Recapturing journalistic authority. *Online Journalism, 7*(1), 65–83.

Rosen, J. (1991). Making journalism More more public. *Communication, 12*(4), 267–284.

Tarde, G. ([1898] 1969). Opinion and conversation. In T. Clark (Ed.), *On Communication and Social Influence.* Chicago: University of Chicago Press.

Vaisman, C. L. (2006, November). Design & play: Weblog genres of adolescent girls in Israel [Electronic version]. *Reconstruction: Studies in Contemporary Culture, 6*(4). Retrieved 2 December 2006 from http://reconstruction.eserver.org/064/vaisman.shtml

Wright, C. (1960). Functional analysis and mass communication. *Public Opinion Quarterly, 23,* 605–620.

CHAPTER EIGHT

Offline Politics
IN THE Arab Blogosphere

Trends and Prospects in Morocco

AZIZ DOUAI

Introduction

Blogging, the ultimate symptom and product of the global proliferation of information, has demonstrated its capacity to bolster an intricate Web of global consciousness and subvert traditional means of newsgathering and dissemination. In the Arab World, the phenomenon has taken similar shades. Blogging has been hailed as a new tool of political dissent that embodies recent calls for genuine broad reform in the region (Rahimi, 2003). Skalli (2006: 51) concludes that information "technology promises to be enabling and empowering educated women in politically and religiously constrained environments." In a pattern reminiscent of the rise of pan-Arab satellite television (El-Nawawy & Iskandar, 2002), Arab bloggers' ascendancy has abetted a growing propensity to ascribe immense political influence to mediated speech (Lynch, 2006). The pattern has whetted the expectation that the Arab blogosphere may herald comparable shifts in Arab politics and public discourse. For a region deemed on the verge of disaster by the United Nations Development Programme (UNDP, 2002), blogging has been embraced as a latest fad to promise political change if not broad social and cultural transformation.

The programmatic agendas of reputable international organizations, such as Reporters Without Borders (RSF) and similar press watchdogs, have followed suit in providing international support and lionizing bloggers worldwide. Two interrelated arguments have been promulgated for such a status. First, bloggers effectively extend the boundaries of free speech and expression enshrined in liberal democracies of the West (Pain, 2004). The second argument draws on Habermas's concept of the public sphere to claim that bloggers actually contribute to the construction of a global public sphere (Hofheinz, 2005). Arab citizens, according to the same argument, are no different in seeking to highlight domestic issues by lobbying a global community's support for their cause(s). Both arguments fail to disguise their optimism about the possibility of change and political reform to be exercised through the Internet and political blogs in particular. The risk that these hopeful notes about blogging run is usually difficult to discern and disentangle unless one remembers cyberspace's tendency to be construed as a mythic space, as Mosco and Foster (2001) argue. As a mythic space, the Internet initially celebrated the end of geography and politics, but now blogs' enthusiasts might unwarily celebrate a different myth, the end of "on-the-ground" activism and "offline" politics.

This chapter examines the ongoing and mixed legacy of the Arab political blogosphere, its potential to enhance oppositional politics, to breed a new culture of dissent, and to invigorate the debate about the Arab and global public spheres. In evaluating this new space, this study steps out of the star-struck gaze at political blogs as full-fledged alternatives to "offline" politics and activist dissent in Arab political culture. A brief overview of the landscape of the Arab blogosphere makes evident how it emerged partially from the geopolitical conditions created by the U.S. intervention, the decimation of Iraqi indigenous media, and the rise of a distinct blogging realm in Iraq. The focus on the experience of Moroccan bloggers teases out the similarities and the differences among diverse national experiences in the Arab World. It indicates the perils surrounding blithe generalizations about the Arab blogosphere and points to a new tension in the emerging primacy of local over pan-Arab discourse. Blogging as a phenomenon would be able to actualize its political promise only in conjunction with a coterminous solid social movement that mobilizes larger swaths of Arab public opinion, this chapter will argue.

The Political Genesis of the Blogosphere

Political upheavals have to be credited for popularizing blogs and transforming them from an esoteric form to a vocal commentator on the events gripping both national imaginations and political agendas. In the United States, for instance, the events of 9/11 constituted such a watershed moment for bloggers' popularity and their infusion into American political discourse. By all accounts, the tipping point occurred during the U.S. presidential elections in 2004, and the controversy generated by forged documents disputing President Bush's military service shown on CBS News.

Bloggers' role in unearthing those forgeries catapulted Dan Rather out of his news anchor chair at CBS News and drew mainstream media's coverage of the blogosphere. Under the operating principle of "join them if you can't beat them," mainstream media have urged their journalists and reporters to establish their own blogs. The blogging Web pages in *The New York Times, The Washington Post, The Guardian,* the BBC and CNN amply illustrate how blogging has infiltrated traditional mainstream media.

The United States' invasion of Iraq, with its ensuing chaos, constitutes a similar momentous event defining the Arab blogger's destiny and worldwide recognition. In the Arab World, the early whispers of this new online expression gained momentum immediately in the post-American invasion of Iraq through presenting an intimate outlook on the consequences of the invasion on ordinary Iraqi lives. Emblematic of the rise of an Arab blogging community, the Iraqi blog Salam Pax became a *cause célèbre* that enticed venerable Western media, such as the BBC and *The Guardian,* to invite the blog's author to contribute columns and articles, commenting on the situation in Iraq (El Khazen, 2006). Although critical of terrorist acts in Iraq, Salam Pax supported the overthrow of the former regime. Other Iraqi blogs gained momentum and recognition, such as the award-winning blog Riverbend, subsequently published as a bestseller *Baghdad Burning,* notwithstanding the fact that other Iraqi blogs were accused of being stooges of the American occupation (Boxer, 2005).

Lynch (2007: 9) identifies three modes Arab bloggers engage in: activists, bridge blogging, and public-sphere bloggers. Lynch explains that

> *Activists* are directly involved in political movements, using blogs to coordinate political action, spread information, and magnify the impact of contentious politics. *Bridge bloggers* primarily address Western audiences, usually writing in English with the intention of explaining their societies. Finally, *public-sphere bloggers* tend to not be directly involved in a political movement, but are deeply engaged with public arguments about domestic (and often Arab or Islamic) politics.

Although it is important to remember that, as Lynch points out, these categories remain fluid, political Arab blogs often exhibit these traits as implicit rationales for their existence on the Web.

In the proliferation of thousands of Arab blogs, called *mudawana,* the formidable challenge becomes which blogs deserve attention as legitimate case studies for academic research. To present an overview of the Arab blogosphere, this chapter has followed a limited number of those blogs anointed by global media and meta-blogs, award-winning blogs, as well as those recognized by global NGOs. Table 8.1 presents a short list of these Arab blogs including: Mahmood's Den (Bahrain), Baheyya (Egypt), Salam Pax (Iraq), Where Is Raed? (Iraq), Riverbend (Iraq), Jar el Kamar

(Egypt), Sand Monkey (Egypt), and The Sabah Blog (Jordan). From Morocco, this chapter has chosen to look at Larbi.org, The View from Fez, Kaoutar's Big World Learner, Lachyab's blog, and Sanaa's blog (see Table 8.1). These popular blogs represent a mixture of Arabic, French, and English language blogs, written primarily by Arab bloggers residing in the Arab world. The analysis investigates how these blogs have positioned themselves as conversations that forge ties with the global public sphere, reviewing both their global implications and the prospects of offline politics in the era of the blogosphere.

Table 8.1. List of Arab blogs included in the study

Blog	Web address	Country of origin
Mahmoud's Den	http://mahmood.tv/?page_id=2	Bahrain
Baheyya	http://baheyya.blogspot.com/	Egypt
Jar el Kamar	http://jarelkamar.manalaa.net	
Sand Monkey	http://egyptiansandmonkey.blogspot.com	
Where Is Raed?	http://dear_raed.blogspot.com/	Iraq
Baghdad Burning (Riverbend)	http://riverbendblog.blogspot.com/	
Iraqi Atheist	http://iraqiatheist.blogspot.com/	
Sabah Blog	http://www.sabbah.biz/mt/	Jordan
Kaoutar's Big World Learner	http://bigworldlearner.blogspot.com	Morocco
Lachyab	http://lachyab.jeeran.com	
Larbi	http://www.larbi.org	
Sanaa	http://sanaa.blogspirit.com	
The View from Fez	http://riadzany.blogspot.com	

Cyberspace Encounters The Global Public Sphere

That blogging enables Arab citizens to participate in the global public sphere finds resonance in both Arab cultural traditions and a mediated Arab public sphere, forged with the aid of new communication technologies. The hallmark feature of Arab political blogs is their intense consciousness of their turbulent public sphere, drawing largely on traditional local styles of narration and connection with the global community. Intent on connecting with global discourse to exhort outsiders' solidarity and contest the legitimacy of the governing regimes, political blogs in the Arab World remain rooted in domestic culture and political reality.

In examining the new public sphere in the Arab World, Lynch (2006: 29) observes a new surge of Arab publics' cognizance of the global or international public sphere, coupled with a collective emphasis on public debate through new mass mediated voices. The new global public sphere includes the construction of a transnational Arab public, whether in the Arab world or in diasporic locations and aggregations, that seeks to escape the traditional repression of the Arab state and invite new shifts in the political landscape. The facilitation of this transformation has mostly been carried out on the beams of satellite television. Due to the contentious claims that Habermas's public sphere as a concept conjures up, some political scientists prefer to limit the public sphere to the dominance of "public arguments" regarding sociopolitical issues facing citizens in the Arab world (Lynch, 2006: 30–32). The "public argument" essential to this conceptualization of the Arab public sphere reflects other prerequisites, mainly interactivity and participation in debating relevant issues among newly empowered audiences. Again, Lynch and those who similarly theorize the "transformation of the Arab public sphere" take their cues from the new pan-Arab satellite television and other traditional mass media. Satellite television stations, such as the Qatar-based Al Jazeera television, constitute the epicenter of such transformation in hosting and urging public arguments, potentially pushing them from the media realm onto the Arab Street at large. At any rate, others who examine the blog phenomenon resort to similar theorization to posit that the Internet through blogging is carrying the day and expanding that role. Bloggers and their ilk extend the "public" nature of argument and debate, with forays into new forms of cyber expression.

Arab bloggers have been likened to new missionaries in their zealous belief in the power of the medium to extend the boundaries of free speech and enhance the prospects of a solid civil society (Hofheinz, 2005). As a Moroccan sociologist, Mernissi's (2003) perspective on how Arab users of information and communication technologies draw on their long traditions of encountering the outside world remains instructive. Mernissi (2003) contextualized the information revolution in the Arab world as a hopeful rise of peripheral voices, women and the youth, who follow in the tradition of "Sindbad" (Sinbad), zapping in their satellite television channels and browsing through Web information and virtual pages that dent the stronghold of the sacrosanct. Sinbad's roots in the Arab folklore, as an adventurer from *The Arabian Nights* (or *A Thousand and One Nights*) welcomes meeting strangers be they human beings, or sea monsters, during his sea adventures. His mythic adventures bespeak of the ancient tradition of encountering strangers as a process of acculturation and intellectual or experiential edification. Mernissi appraises Sinbad's fortunes and status thus:

> The only thing predictable in the seven trips Sindbad undertook was that to communicate with strangers, be they humans or even birds and sea-monsters, made him richer and happier. Communication with the stranger who manifests God's cosmic capacity to manifest itself in diverse images is one of the strong messages which runs through Sindbad's tales and explains why, among all the "1001 Nights" stories, only this one became a universal heritage, enchanting European and American children alike.

The Sinbad tradition permits the reconstruction of a global discourse undaunted by, and in fact welcoming of, strangers. It constructs local communities as part of a larger community to draw on for stories, adventures, and global solidarity.

Arab bloggers resemble this ancient seafarer in being adventurous and conscious of the other, the other as an opportunity for an interesting encounter rather than a problematic issue. They welcome debate and discussion of any topic—politics, religion, and relations between the sexes. In fact, these historical taboos, traditionally relegated to the private sphere since their discussion had always been fraught with risks, now furnish the main discussion diet of Arab blogs. In Bahrain, for instance, Mahmood's Den is among the most well-known and respected bloggers that have commented on the state of freedom in his Gulf state. Mahmood's Den unfailingly voices displeasure with the government's unceasing harassment and restrictions on free expression. The increasing popularity of Bahraini bloggers induced Bahrain's Ministry of Information to require bloggers to register with it, a step that many bloggers called an outright infringement on their freedom. In 2005, Mahmood Al-Yousif, the author of Mahmood's Den, did not hesitate to express his outrage, contacted Western journalists about the new edict, and pointedly proclaimed on his Web site "I don't need the MoI's [Ministry of Information's] protection, thanks very much!" (Glaser, 2005, n.p.). Featuring political and cultural commentary on Bahraini and Arab affairs, Mahmood's Den proved to be so irritating to the Bahraini government that they decided to shut the blog down and block it for an extended period. The Ministry of Information's blockade was eventually lifted in February 2006 only after a much publicized campaign that enlisted the support of press watchdogs, human rights' activists, and bloggers from around the globe. The vibrancy of Bahrain's blogosphere was also tested in Bahraineonline's blogger, Abdel Imam, and his collaborators who were detained by the authorities for "violating the press, communications, and penal codes" (according to the Committee to Protect Journalists' 2005 report). News of the arrest was broken to the world through fellow bloggers such as Mahmood's Den and Chan'ad Bahraini, causing an outcry from Reporters Without Borders and other media watchdogs (RSF, 2005a). The worldwide publicity of the arrest generated by the blogosphere took action into the street through demonstrations, leading to the release of the detained Bahraineonline bloggers although the charges against them were not immediately dropped.

However, the Arab blogosphere will not resemble a democratic and fully representative public sphere without encouraging women to take part in the new opportunities. Although blogging grants anonymity and safety for Arab women to venture into a male-dominated mediated terrain, it seems that only a few female Arab bloggers have partaken of the new opportunity. From Egypt comes the female blogger "Baheyya: Egypt Analysis and Whimsy." With a nom de guerre standing for Egypt, her blog offers "commentary on Egyptian politics and culture by an Egyptian citizen with a room of her own." Baheyya's posts stem from a reformist perspective that condemns the country's authoritarian regime while celebrating its progressive intellectuals from Shaykh Imam Eissa to Ismail Sabri Abdallah (November 8, 2006). Another example of how Arab women have taken to blogging is Baghdad Burning by an Iraqi woman, a blog that has made headlines among international media (the next section provides further details on this blog). Overall, Arab women have used their blogs as a platform that builds on the Internet's capacity to enable "virtual activism" despite patriarchal conditions inhibiting their participation (Tadros, 2005). For most Arab bloggers, then, cyberspace encounters turn into an excuse to disseminate information about Arab affairs, culture, and politics. The influence of politics, global violence, and war on the Arab blogosphere cannot be overstated.

Blogging the war and the State

Not all encounters of Arab citizens with the outside world, however, have been peaceful for the memories of European colonialism and foreign interventions run deep through the Arab mind. The recent U.S. intervention in Iraq, the latest bidder in these imperialist encounters, constitutes a tipping point that has energized Arab bloggers, especially Iraqi blogs, and led to their worldwide recognition. The Iraq Blog Count, a "meta-blog" that tracks both Iraqi bloggers and blogs about Iraq, identifies at least about 477 Iraqi blogs in Arabic also aggregated on another Web site (maktoobblog.com). In addition to their evident focus on the war, one Iraqi blogger sheds light on the fractured politics of religion. Right from the get go, the blogger labels her/himself "Iraqi Atheist" and revels in an indictment of religious infighting that is crippling the country. From Baghdad and "amidst all the religious hysterical mess in Iraq," the "Iraqi Atheist" promises the blog reader, "I will try to set a standard of an alternative way of thinking and living." A raw outlook on the internecine strife remains the hallmark of this blog, portraying the ethnic patchwork of Iraqi Arabs and Kurds as well as the treacherous religious map of a land unevenly spread out among Sunni, Shiite, and Christian religious factions.

"Where Is Raed?" is among the best-known Iraqi blogs that documented the final months preceding the invasion and the chaotic aftermath of the occupation. Under the pseudonym of Salam Pax, the blogger did not shy away from renouncing the violence unleashed in Iraq and obliquely linked it to an imperialist venture. Salam Pax utilized a Huntington's quote to drive this indictment home, placing it right at the top of the left column of his blog: "The West won the world not by the superiority of its ideas or values or religion but rather by its superiority in applying organized violence. Westerners often forget this fact, non-Westerners never do." Indeed, the bulk of his blog uneasily draws a cautious optimism at the fall of the former Baathist regime, only to swerve to a final pessimism in a penultimate post on April 6, 2004. The penultimate post of Salam Pax ominously assesses the vaunted freedom in Iraq by a leading comparison: "Remember the days when every time you hear an Iraqi talk on TV you had to remember that they are talking with a Mukhabarat [Saddam's secret police] minder looking at them noting every word? We are back to that place." He was referring to the surge of a different sort of freedom muzzling politicians, the new religious fundamentalists of the Shiite Al Sadr faction that the occupation has enabled into existence.

"Baghdad Burning" is an Iraqi girl's blog that challenges the reader from the outset describing her blog in her opening statement as "Girl Blog from Iraq . . . let's talk war, politics and occupation." The occupation aside, she vents her searing outrage at the postinvasion government in Iraq, its inability to restore peace and order, and its tendency to restrict freedom or incite sectarian violence. In November 5, 2006, upon the breaking news of Saddam's guilty verdict and his death sentence, the blogger cynically opines: "When All Else Fails . . . Execute the Dictator." She details how pro-Saddam demonstrators have been clamped down and goes on to comment on the political process:

> I'm more than a little worried. This is Bush's final card. The elections came and went and a group of extremists and thieves were put into power (no, no—I meant in Baghdad, not Washington). The constitution which seems to have drowned in the river of Iraqi blood since its elections has been forgotten. It is only dug up when one of the Puppets wants to break apart the country. Reconstruction is an aspiration from another lifetime: I swear we no longer want buildings and bridges, security and an undivided Iraq are more than enough. Things must be deteriorating beyond imagination if Bush needs to use the 'Execute the Dictator' card.

> Iraq has not been this bad in decades. The occupation is a failure. The various pro-American, pro-Iranian Iraqi governments are failures. The new Iraqi army is a deadly joke. Is it really time to turn Saddam into a martyr? Things are so bad that even pro-occupation Iraqis are going back on their initial 'WE LOVE AMERICA' frenzy (Riverbend/"Baghdad Burning," November 5, 2006)

Even new Iraqi independent news channels, such as Zawra and Salahiddin television networks, have been raided by security forces and eventually shut down. "Baghdad Burning" remains primarily devoted to political commentary, pondering the abyss while being on the verge of declaring the utter failure of a new Iraq. Focus on ethnic cleansing and internecine violence is clearly the blog's hallmark.

Egyptian bloggers have similarly pursued the mantle of political dissent that challenges both their authoritarian regime and the rise of fundamentalist politics in the Muslim world. Estimates of the number of Egyptian bloggers vary widely, putting them well into the hundreds, if not thousands, and the best-known Egyptian blogs are in English as well (Levinson, 2005). There is no attempt to disguise the politics of Egyptian bloggers, if not actually celebrate it in earnest on their Weblogs. For instance, Egyptian Sand Monkey blog invites its readers with a cynical notice: "Be forewarned," the short bio clip quips, "The writer of this blog is an extremely cynical, snarky, pro-US, secular, libertarian, disgruntled sandmonkey. If this is your cup of tea, please enjoy your stay here. If not, please sod off." And the various posts on the Sand Monkey blog do not disappoint a reader intent on identifying those politics of dissent. From the first posts streaming in December 2004 to the present, Sand Monkey revels in demonizing Islamists, delving into U.S. politics, pop culture, and the state of Arab music videos, and cruising into the politics of Israel's war on Lebanon. The "rantings" of the Sand Monkey are flippant; yet, they offer a blistering criticism of Islamists. For those who might characterize the blog as Islamophobic, the blogger has this to say: "I am not taking every opportunity to criticize Islam, nor am I an Islamophobe . . . I am just pointing out the issues that I think deserve to be debated in our Islamic culture and I welcome anyone's opinion who can actually give me an argument or show me where I am wrong. And no, 'Cause God said so' doesn't cut it!" (April 28, 2005). Egyptian mainstream media deserve total disdain for they are nothing but "a mouthpiece for Mubarak," declared Sand Monkey (April 26, 2005).

The Deutsche Welle's coveted prize, "Best of the Blogs" Award went to an Arabic language blog from Egypt, Jar el Kamar (The Moon's Neighbor) in 2006. In its rationale for the award, Deutsche Welle describes this blog in the following manner: "An example of citizen journalism, Jar el Kamar has been able to cover incidents in his local city of Alexandria more bravely than your typical media outlet. He was on the front line of dangerous situations including violence happening around parliamentary elections and church attacks. He also blogs about culture, social issues and other topics" (The BOBs, 2006). The coveted prize represents one means of both institutionalizing and legitimizing the Egyptian blogosphere. The blog distinguishes itself for not dabbling too much into the political debates dominating the local blogosphere although the blogger remains undoubtedly critical of the Egyptian authorities' repression. In November 2006, Jar el Kamar published an update on Egypt's crackdown on cyberspace freedom and the arrest of Kareem Amer, a fellow blogger, for the second time, and drawing attention to a Web site (www.freekareem.org) set up to demand his release. Overall, Jar el Kamar sounds more authentic and genuine an Egyptian voice than many blogs from Egypt.

To continue sampling the Arab blogosphere, Haitham Sabbah is a well-known blogger from Jordan whose Palestinian roots are revealed in his frequent posts on Israel and Palestinian affairs. Mr. Sabbah's blog finds motivation in his "disappointment of [sic] what [he] read[s] about what and how the Middle East is represented to the world online," as the blogger explains in his brief bio. His "chief interest is in the intersection between politics and individual liberty in the Middle East and Muslims [sic] world." The focus on individual liberties in the Middle East includes publicizing evidence of Egypt's state torture (December 1, 2006 post), a blistering critique of Human Rights Watch's assessment of Palestinian violence (December 2, 2006), and the growing numbers of Israeli settlements in the occupied territories (December 6, 2006). In sum, the Arab blogosphere is very much entangled in the political arena. A careful analysis indicates a rising tension associated with the preeminence of the local, as demonstrated by the following examination of the Moroccan blogging community or "blogma."

In "Blogma," Blog Locally

The analysis of the content of the Arab blogosphere has so far rooted its genesis and content in the political upheavals taking Arab societies by storm. However, such analysis does not seek to obfuscate the fact that the Arab blogosphere remains a fragmented space, such as the Web itself, with a politics difficult to discern. Providing national cases from different Arab states indeed illuminates some of the difficulties that generalizations regarding an "Arab blogosphere" may engender. In other terms, the Arab blogosphere tends to reflect both fragmentation and pluralism that parallel the mediascape of pan-Arab media. Unlike traditional pan-Arab media, however, blogging about local issues constitutes a rising trend characterizing this new space. Lynch (2007) observes that the Arab blogosphere can present an alternative space that gives primacy to local issues, rather than pan-Arab issues, and thus aid in constructing genuinely representative and deliberative public discourse fora. Such a claim can find its best support in the special case of the Moroccan blogosphere, a nascent experience compared to other "established" blogging spaces in the Arab world. First of all, the Moroccan blogosphere remains dominated by French language blogs, unlike other Arab bloggers in the Middle East who prefer to use English as their language of choice. In terms of their proliferation, an estimate of Moroccan blogs puts their number in the neighborhood of a thousand, according to "Annuaire des blogs marocains," a Web aggregator of Moroccan blogs.

Blogma, or the Moroccan blogosphere, has undergone a transformation resembling other national experiences in the Arab blogosphere. Although Moroccan bloggers have not run afoul of the authorities yet, they have plunged headlong in the tumultuous political debates raging in the "Arab Street," albeit with an important difference, privileging the local over pan-Arab issues. A cursory examination of "blogma" leads to the impression that local politics and domestic concerns dominate online debates among Moroccan bloggers. Farah Kinani, a frequent contributor to Harvard University's blog aggregator Global Voices Online, made a similar observation when she noted that "The Moroccan blogosphere barely acknowledges Saddam Hussein's sentence" (Kinani, 2006, n.p.). Instead of debating the significance of the execution, the Moroccan blogosphere brushed it aside and focused on the question of the Moroccan Sahara, and a first meeting of Moroccan bloggers in the city of Agadir, among other domestic issues.

No wonder that freedom of the press and speech stand out as hot-button issues in blogma. Upon the publication of some jokes that were deemed hostile or insulting to Islam and the monarchy in December 2006, *Nichane*, a Moroccan weekly, faced legal action from the Moroccan government, with both its editor and one of its reporters facing a jail sentence of three to five years. Moroccan bloggers soon seized on the story, declaring their "solidarity" with *Nichane*. Larbi of the eponymous Larbi.org, a well-known French language Moroccan blog, considers the jokes' topics, sex, religion, and politics, "a dangerous combination" in Morocco, but avows his support for freedom of speech against fundamentalists and totalitarianism. The trend among the 454 reader posts in response to the blogger's article is overwhelming in its support of the magazine. The English language Moroccan bloggers in "The View from Fez" also highlighted the charged atmosphere that the jokes have led to, and how Moroccan Islamists and religious scholars misappropriated the issue to crack down on free speech.

In the wake of the terrorist attacks in Casablanca in March and April 2007, the Moroccan blogging community's response was predominantly utter condemnation and horror. The "Big World Learner," Kaoutar's English language blog from Rabat, voiced the opinions of many when she sought to distance Islam from the terrorists who blew themselves up:

> If they imagine they are waging a "holy" war, the Islam I know prohibits even the destruction of plants during war time!! Let alone innocent people!

> The Islam I know prohibits killing women and children during war time!! So on what basis are they making random carnages?!

The comments posted on her entry concurred as much, with some calling for a re-examination of the motives and the root causes of terrorism. Other bloggers and

Web surfers made their sentiments and responses public on the Web. Larbi's post entitled "Injustifiable terrorisme" ["Unjustifiable Terrorism"] drew 134 posted comments, vibrantly debating the problem and the solutions, but all condemning the act. The picture of unanimous condemnation would be perfect but for the Arabic language blogger's Lachyab.jeeran.com, which exposes more diverse views, such as those which claimed that the attacks were a government's conspiracy to sully the reputation of Moroccan Islamist parties. In short, even a snap look at Moroccan bloggers' reflections on the terrorist attacks in Casablanca provides a sense of the public, and sometimes not so public, discourse on the issue.

Skalli (2006) argues that Arab women are increasingly invading the male-dominated media landscape. The risk of providing only a normative nod, or "only a passing recognition" (Skalli, 2006: 36), to this important aspect of the Moroccan blogosphere exists. Nevertheless, Moroccan women bloggers have appropriated the new medium in proportions equal to men. Their posts vary widely from the personally confessional to "serious" politics. A vibrant discussion ensues on Sanaa Elaji's blog, a journalist and a blogger, who was sued by the Moroccan government for publishing those "unseemly" jokes insulting Islam in *Nichane*. Although most posts declare their solidarity, some demand her "tawba," a declaration of culpability, expression of penitence, and request for forgiveness. Some comments on her blog even border on physical threat. Ms. Elaji's blog may be an extreme case, but it serves to highlight the unprecedented outspokenness of female voices in blogma. *Nichane*'s and Elaji's travails ended with the government's dropping their charges. The rising female voices in blogma and the dominance of local, instead of pan-Arab, issues constitute its strongest features.

The dominance of local issues in the topics that Moroccan bloggers tackle is a development that does not necessarily herald an utter severance of their ties with the larger Arab blogosphere or pan-Arab issues. Severing such ties remains impossible because blogging is fundamentally about establishing transnational networks of solidarity and support. To air domestic corruption and the political dirty laundry of Arab regimes among larger audiences and, in fact, induce international condemnation of a government's crackdown on freedom of speech for instance, constitutes the *modus operandi* of most Arab political bloggers. As mentioned earlier, the support garnered for "Mahmood's Den" and the "Free Kareem" campaign hints at the global reach and effectiveness of these solidarity networks. Moroccan bloggers have connected with this transnational movement, and they participate in it both as Arab bloggers and as members of the global (blogging) community. Hence, it is difficult to deduce from the flow of local issues, or the ebb of pan-Arab discourse, in blogma the death of pan-Arab mediated discourse. What the Arab blogosphere does is initiate a successful counter-balance to the mainstream pan-Arab media's negligence of the local in their hot pursuit of transnational Arab audiences.

Arab bloggers are thus strategically using the paradigm of traditional community media, such as local radio's and newspapers' focus on local communities and issues, to challenge the dominance of pan-Arab issues over their local affairs. As bridge builders, bloggers are burying the hatchet by reconciling the local with the pan-Arab rather than spoil for a fight. Thanks to these new "missionaries," the outcome, the unique mélange of the local, the pan-Arab, and the global in the Arab blogosphere, remains possible.

Blogging and Offline Politics: Myths Reconsidered

The diverse strands of the Arab blogosphere provide a breathtaking case of how the new medium can effectively contribute to a public debate that connects the local with the global and mounts a threat to the status quo. A sober assessment of this new sphere, however, has to admit that Arab bloggers' overall contribution remains a "mixed bag," oscillating between inhibiting and fostering a vibrant political transformation. The oscillation tends to be the result of both the subjective and objective reality of the blogging community. Some hurdles obstructing the full potential of bloggers as local agents of change are indelible from the Arab world's information and human development. Chiefly, language friction, low literacy rates, and the information technology infrastructure within the region are formidable obstacles.

The present chapter reviewed how Arab bloggers have ushered new shifts in the paradigms of news, information flow, and politics inside the Middle East. How these shifts are occurring poses a further question regarding the viability and sustainability of the new medium as a means of political dissent. How does the facilitation of an oppositional culture celebrate virtual politics at the expense of, or as a substitute for, offline politics? What implications do such shifts hold for regional political reforms? Or, are we merely witnessing a new mediated myth that disregards Arab information reality? How do issues of language and literacy affect bloggers' reach? For sure, new technologies have always been associated with new myths and revolutions bordering on a risky hype of their transformative power. And cyberspace has been a breeding ground of political myths that incessantly proclaim the end of politics. To further the thesis posited so far, the contradictory and "mythic" nature of the Arab blogosphere, I turn to Mosco and Foster's notion of cyberspace as a "myth."

Mosco and Foster (2001: 218) argue that cyberspace can be best theorized as a "mythic space" that celebrated "the end of history, the end of geography, and the end of politics." Mosco and others have variably debunked these myths in their ways, arguing that cyberspace will experience the same "enclosures" characterizing capitalism (e.g., Bettig, 1997). With the rise of new forms of Web expression such as blogging and traditional online forums, the danger may no longer be the end of politics but ironically a different sort of myth: the Internet as a powerful political tool that makes human collective action offline an obsolete mode of resistance. At stake here is an unconditional submission that the Internet might substitute for activism on the ground, which smacks of an "anti-politics" vision, one that echoes its sibling "the end of politics" myth.

The new myth does not lack some historical lessons to provide it with legitimacy. The question of the political efficacy of bloggers has hinged on their successful development and connection with indigenous social movements which mobilize for social and political change. Social movements, those that particularly espouse progressive politics, can provide a well-tested terrain for action as well as act as a source of uniting the potentially fragmented political message of blogs. Prolific examples of the positive transformation of the Internet into a political activist tool abound. The Zapatista movement's celebrated use of this medium to broadcast its grievances and challenge the repressive Mexican state's manipulation of mass media is a historic precedent (Best & Kellner, 2001). The anti-globalization protest movements from Seattle to Porto Allegre provide fodder for thought on the political implications of the Internet in fostering global connections. Bloggers from the Arab world indirectly nourish these connotations and evolving Internet history in voicing their dissent. As noted above, the main problem lies in the unsophisticated use of local organizing that might turn them into a collective action network, however-er. Whereas bloggers in the United States have molded their reach into a political arm, Arab bloggers have been unable to translate their cyberspace dissent from online talk into an offline mobilizing force. The challenges thus endure as internal and external, subjective and objective.

Myths cannot transcend reality although they may succeed in obfuscating it for some time. The Arab information reality inhibits the unleashed promise of the Internet, and bloggers in particular, within the Arab world. A quick look at both literacy rates and the penetration of the Internet reveals the existence of informational gaps across citizens and the states of the region. International assessments of the Arab information society provide a sobering reality check. The UNDP 2002 report, for instance, portrayed a region lagging behind in human and informational development and warned that "reform is necessary and if it is to be successful and sustainable then change has to come from within" (Brown, 2003: 4). Other stern messages were reiterated in subsequent reports, notably focusing on "a knowledge deficit" as a "cardinal challenge" preempting any genuine development path (Brown, 2003: 4). Although the Internet has occupied the front lines of knowledge diffusion and information dissemination worldwide, an empirical examination reveals that the prohibitive costs of access, relative to income, handicap its diffusion amongst Arab citizens. There are large variations in the distribution of Internet penetration within Arab countries themselves. Disparities in income and wealth impact access to the technology, mainly personal computers, stymieing the number of Internet users.

Myths may gloss over fractious issues. One such issue in the Arab blogosphere is its language, an intractable source of friction that it needs to confront. Although many Arab bloggers tend to use Arabic, only those who use English, or French as in the case of North African bloggers, are more likely to attract the attention of the global advocacy rights community. The consequence of using a foreign language potentially disenfranchises large sections of Arab citizens who do not have the pre-requisite linguistic skills to join the debate or effect change. It is worthwhile to remember that the first celebrated Arab blogs were transcribed in the English language, as Al-Sabah Blog (Jordan), Baghdad Burning (Iraq), Mahmood's Den (Bahrain), and The View from Fez (Morocco). Language use among Arab bloggers poses the issue of the primary audience they wish to target. Since most Arab citizens do not speak English, or French for that matter, the logical surmise is that the primary audience of Arab blogs in English will be the global English speaking community. Although unnerving for the analyst of the Arab blogosphere, the use of English is not very surprising since the Internet has been predominantly English language tilted. The largest number of Arab blogs remains written in Arabic language, but few of those blogs have gained global recognition as in the case of the Deutsche Welle award-winning Jar el Kamar (Egypt). The language issue varies according to national context, but some Moroccan bloggers have initiated a campaign for the Arabization of Moroccan blogs, a movement that has not gone unreported in traditional news media (Al Jazeera News Channel, 2007).

Finally, myths are paradoxical. The paradoxical nature of the Arab blogosphere illustrates a "mixed bag" of blessings and disguised curses. The blogosphere signals a popular shift toward an incipient "citizen journalism" in a region where press freedom is by no means on the march (RSF, 2005b). Although bloggers tend to be educated and part of the elites, they also report events from angles that depart from the traditional mainstream media in the Arab world. As discussed earlier, the tension in blogma's transcendence of pan-Arab mediated discourse by privileging the local is a direct challenge to trans-Arab media discourse. The ability to circumvent Arab governments' censorship, or the ideological dictates of a partisan press in some cases, enhances the appeal of the blogosphere as an attractive forum for dissent. And that has come at a high price, as shown earlier, a price that includes incarceration and physical threat. According to Reporters Without Borders, an increasing number of bloggers in Egypt and Bahrain have been jailed, sued, and harassed by the authorities. Six out of fifteen governments that RSF classifies as "enemies of the Internet and other countries to watch" in 2005 hail from the Arab World: Libya, Saudi Arabia, Syria, Tunisia, Bahrain, and Egypt (RSF, 2005b). The fear exists that repressive regimes might bundle what they perceive as "troublesome" bloggers with traditional media's calls for more freedom, leading to the repression of both.

References

Al Jazeera News Channel (2007, March 24). "Ansaro lughati al-arabiya bel maghreb yataharrak-ouna li himayatiha" [Defenders of the Arabic language in Morocco are moving to protect it]. Retrieved April 20, 2007, from http://www.aljazeera.net/NR/exeres/BD134001–31A7–4EAF-90E4–69492AA8F1F3.htm

Best, S. and D. Kellner (2001). *The postmodern adventure.* New York and London: Guilford Press and Routledge.

Best of the Blogs (BOBs) (2006). Deutsche Welle's best of the blogs international award 2006. Retrieved from http://www.thebobs.com/index.php?w=1159538015691413LKQRGYPM

Bettig, R. (1997). The enclosure of cyberspace. *Critical Studies in Mass Communication, 14,* 38–157.

Boxer, S. (2005). Pro-American Iraqi blog provokes intrigue and vitriol. *The New York Times.* Retrieved January 18, 2005 from http://www.nytimes.com/2005/01/18/arts/18blog.html?_r=1&oref=login

Brown, M. M. (2003). Foreword. In *Arab human development report 2003: Building a knowledge society.* New York: United Nations Development Program. Retrieved in March 30, 2006 from http://hdr.undp.org/en/reports/regionalreports/arabstates/arab_states_2003_en.pdf

Committee to Protect Journalists (2005). *Attacks on the press: A worldwide survey.* Retrieved February 23, 2006 from http://www.cpj.org/attacks05/mideast05/snaps_mideast_05.html#-bah

El Khazen, J. (2006). Ayoon wa Azaan: Arab blogs. *Al Hayat.* Retrieved January 14, 2007 from http://english.daralhayat.com/opinion/OPED/01–2006/Article-20060117-d8ed5b3b-c0a8–10ed-0013–5f0aaccda524/story.html

El-Nawawy, M., & Iskandar, A. (2002). *Al Jazeera: How the free Arab news network scooped the world and changed the Middle East.* Boulder, CO: Westview Press.

Glaser, M. (2005, May). Online forums, bloggers become vital media outlets in Bahrain. USC Annenberg's *Online Journalism Review,* May 2005. Retrieved March 25, 2006 from http://www.ojr.org/ojr/stories/050517glaser/

Hofheinz, A. (2005, March). The internet in the Arab world: Playground for political liberaliza-tion. *Internationale Politik und Gesellschaft/International Politics and Society,* pp. 78–96. Retrieved April 16, 2006 from http://fesportal.fes.de/pls/portal30/docs/FOLDER/IPG/IPG3_2005/07HOFHEINZ.PDF

Kinani, F. (2006). The Moroccan blogosphere barely acknowledges Saddam Hussein's sentence. Retrieved January 12, 2007 from http://www.globalvoicesonline.org/2006/11/08/the-moroccan-blogosphere-barely-acknowledges-saddam-husseins-sentence

Levinson, C. (2005, August 24). Egypt's growing blogger community pushes limit of dissent. *The Christian Science Monitor.* Retrieved February 27, 2006 from http://www.csmonitor.com/2005/0824/p07s01-wome.html

Lynch, M. (2006). *Voices of the new Arab public: Iraq, al-Jazeera, and Middle East politics today.* New York: Columbia University Press.

Lynch, M. (2007, February). Blogging the new Arab public. *Arab Media & Society, 1,* 1–30. Retrieved March 15, 2007 from http://www.arabmediasociety.com/?article=10

Mernissi, F. (2003). The Cowboy or Sindbad—Who will be the globalization winner? Retrieved February 25, 2006 from http://www.mernissi.net/books/articles/sindbad.html

Mosco, V., & Foster, D. (2001). Cyberspace and the end of politics. *Journal of Communication Inquiry, 25*(3), 218–236.

Pain, J. (2004). Bloggers, the new heralds of free expression. In *Handbook for bloggers and cyber-dissidents*. Paris: Reporters Without Borders. Retrieved April 14, 2005 from http://www.rsf.org/article.php3?id_article=14998

Rahimi, B. (2003). Cyberdissent: The internet in revolutionary Iran. *Middle East Review of International Affairs, 7* (3), 101–115.

RSF (2005a, February 28). Three Barhainonline.org moderators freed [Press release]. *Reporters Without Borders*. Retrieved March 15, 2006 from http://www.rsf.org/article.php3?id_article=12687

RSF (2005b, November 17). The 15 enemies of the internet and other countries to watch [Annual report]. *Reporters Without Borders*. Retrieved June 12, 2006 from http://www.rsf.org/article.php3?id_article=15613

Skalli, L. H. (2006). Communicating gender in the public sphere: women and information technologies in the MENA. *Journal of Middle East Women's Studies, 2*(2), 35–59.

Tadros, M. (2005, March). *Arab women, the internet, and the public sphere*. Paper prepared for the Mediterranean Social and Political Research Meeting, Florence, Italy. Retrieved January 26, 2007 from http://www.mengos.net/books/reports/TadrosMedMeetpaper.pdf

United Nations Development Programme (UNDP) (2002). *Arab human development report 2002: Creating opportunities for future generations*. Retrieved July 8, 2006 from http://www.fimam.org/Informe%20PNUD%202002CompleteEnglish.pdf

CHAPTER NINE

Beppegrillo.it

One Year in the Life of an Italian Blog

GIOVANNI NAVARRIA

Beppe Grillo: The Talking Cricket

In Carlo Collodi's classic children tale, *The Adventures of Pinocchio*, a talking cricket (*grillo* in Italian) is killed by Pinocchio for trying to impart wisdom to the wooden-headed marionette. In the contemporary Italian media landscape there is another controversial cricket, Beppe Grillo, one of the most popular and controversial stand-up comedians who has ever appeared on Italian television. Grillo began his career at the end of the 1970s (Internazionale, n.d.) and by the early 1980s, high audience ratings and critical acclaim made him a national celebrity. Toward the end of the decade, he began criticizing prominent Italian politicians and big corporations for corrupt practices (Grasso, 1992: 467–468; Israely, 2005). Because of mounting pressure of politicians and advertisers against Grillo's satire, TV producers stopped inviting him on their shows. Sent into unofficial exile, Grillo was forced to perform in theatres, sports arenas, and public squares.

Since the early 1990s Grillo has appeared only twice on public television. Yet Grillo's ban from the small screen has made him even more popular with the Italian public (Internazionale, n.d.), which regards him as the outspoken talking cricket, a vociferous critic of political and economic corruption. In 2005 *Time* magazine named Grillo among the 37 European heroes of the year (Israely, 2005).[1] In recent times, Grillo has been able to increase his popularity by transforming himself from a popular television comedian into a blogger. Through his site beppegrillo.it Grillo and his staff offer nonaligned and critical political information that rarely finds space in today's mainstream media. At the same time, thanks to the comments and countless feedbacks that are either posted daily on the blog or sent via email, Grillo himself has access to information and stories that otherwise would remain untold.

In this chapter, I argue that beppegrillo.it represents both the dark and the bright sides of grassroots politics on the Web. Digital tools and networks offer potential new ways to facilitate political engagement; they can also serve to undermine the democratic process they champion by adopting a double standards approach on three important pillars of any democratic endeavor: transparency, accountability, and representativeness.

On the one hand, democratic institutions, elected representatives, watchdog and civil society organizations are all important elements of the complex mechanism of a democracy, and they are all—in various degrees—expected to uphold these three fundamental pillars (Held, 1996). On the other hand, political formations facilitated through digital tools, such as blogs, do not require these three elements. Blogs like beppegrillo.it, attract thousands of readers every day and can act as a point of reference in important political debates. The more political they become the more they should consider whom if anyone they represent and to what extent they should be accountable and transparent. Often, however, bloggers avoid these questions because of the ambiguous role they play in public sphere: Are they journalists? Are they political subjects? Or are they merely the solo voice of a chorus of angry citizens?

Grillo is a vociferous critic of the lack of democratic openness in contemporary Italian politics. He fights to unveil the truth about issues that mainstream media and politicians do not dare to address (Grillo, 2004: 405). His main arguments can be summarised in three lines of critique: (1) politicians (and also high ranking civil servants) should be held accountable for their actions (Grillo, 2006d); (2) to be truly representative they should be chosen by the people and not by political parties, as is often the case in Italy (Grillo, 2007a; Povoledo, 2007); (3) politicians and their actions should be fully transparent (Grillo, 2007a). As Grillo's actions are becoming increasingly political and action oriented, however, Grillo needs to confront the questions of accountability, representativeness, and transparency in order to strengthen its political potential and protect against accusations of shallow demagoguery.

The case of beppegrillo.it also represents an important example of how civic-minded people with limited access to mainstream media, but who are equipped with a strong sense of civic engagement and a history of integrity and who are willing to support others, can indeed harness the power of the Web to promote innovative modes of political participation. Overall, beppegrillo.it is more than simply a blog, it functions as an electronic beacon whose signals manage to attract on its virtual shores an otherwise fragmented and geographically dispersed public. The texts of the many thousand comments, published on the blog and from the quick and continuing blossoming of many hundreds of Friends of Beppe Grillo Meetup groups, demonstrate that this virtual space is home to a flourishing community of individuals looking to fight against the establishment.

The political background in which the blog was born and bred has clearly been a key factor in its expansion. The growing shadow of the media regime established by Silvio Berlusconi, while serving as Italy's prime minister (2001–2006), has on one hand muted criticism coming from mainstream media. On the other hand, it has produced two unintended consequences: (1) the Internet has virtually remained untouched by censorship, (2) the silencing of mainstream media has pushed non-aligned audiences toward new alternative sources of information such as beppegrillo.it. Thus, it is not surprising that the comedian's blog, in a short period of time, has become one of the main reference points through which many Italians, scattered around the country and across the globe, can make sense of the state of things in the country.

Berlusconi's Media Regime

Silvio Berlusconi is the richest man in Italy (*Forbes Magazine*, 2007) and the owner of the largest commercial television group, *Mediaset*, through which he personally controls three country-wide television networks (*Canale 5, Italia 1, Rete 4*). The late Indro Montanelli was a strong critic of Berlusconi's power and one of the most respected Italian journalists of the twentieth century. For Montanelli, Berlusconi with his predominant position in the Italian media landscape represented a great danger for democracy. "Nowadays," Montanelli argued "to introduce a regime, one no longer needs to march towards Rome, nor does one need to set fire to the Reichstag, neither one needs a coup at the Winter Palace. All that is needed are the so-called mass communication media: and among them, sovereign and irresistible is television" (Gomez & Travaglio, 2004: back cover).

Serving as prime minister, between 2001 and 2006, Berlusconi, effectively, was also in control of the Italian public service broadcaster, *Radiotelevisione Italiana* (RAI). Created in 1954, RAI has developed in a complex state-owned media company comprised of three terrestrial nation-wide networks, radios, satellite and Internet television. Its main revenue is based on a national TV license fee and is administered by a nine-member board. By law, these board members are chosen by political parties—seven elected by a parliamentary committee and two by the Ministry of Finance (*La Repubblica*, 2005).

Due to Berlusconi's media monopoly—his strong institutional grip on RAI, his business control of Mediaset,[2] and the silencing of the center-left press (Blatmann, 2003; Gomez & Travaglio, 2004: 217–246) he was able to establish a firm media regime during his five years in power, run by people *willing to* support Berlusconi's own version of truth. Furthermore, as the historian Paul Ginsborg puts it, Berlusconi "has always had his own team of 'organic' intellectuals of variable quality [...] whose programmes have barked out the line incessantly, at all times of the day and night" (Ginsborg, 2003: 38).

One famous example of the influence of the regime on Italian media is the Italy-Germany diplomatic row that took place in the summer of 2003. When in July of that year Berlusconi caused a wave of indignation throughout Europe by comparing a German member of European Parliament, Martin Schultz, to a Nazi concentration camp commander (*The Guardian*, 2003a), RAI's main evening news program did not show the incident and only briefly reported on it; coverage on other stations was "deliberately softened and cut" (Arie, 2003). The Italian press downplayed the affair, "with many papers relegating the story to the back pages" as reported by the BBC (2003).

Between 2001 and 2006, Berlusconi's unique media regime[3] was capable of casting a heavy curtain of silence over information that might have damaged the prime minister's image and business interests. Amid a series of trials and investigations into the sources of Berlusconi's wealth that could have ruined him politically and economically, his control of media muzzled any attempt at thorough analysis of those trials and their revelations.[4] Not surprisingly, in this political milieu, Freedom House listed Italy as the least democratic country in Europe: Italy was ranked eightieth in the world, immediately after Tonga and Botswana and just before Antigua and Burkina Faso (Freedom House, 2006).

The part of the Italian public who disagreed with Berlusconi's politics and his methods resisted by seeking ways to break through the censorship and control that sustained the regime. Traditional means of resistance, such as public gatherings and picketing, are ineffective when television networks refuse to report them (Gomez & Travaglio, 2004: 284–291). For example, during the 2003 campaign against the Iraq War, 3 million people gathered in Rome, yet the protests were not reported by RAI in order to spare politicians pressure from the people (*The Guardian*, 2003b). Roberto Natale, head of RAI Journalists Union, said he and his colleagues at the station were instructed to downplay the size of the protest, not to show the pacifist flag, and to refer to the protesters not as *pacifisti* (pacifists) but with the negative adjective of *disobbedienti* (disobedient people) (Gomez & Travaglio, 2004: 289; *The Prime Minister and the Press*, 2003). However, there are other ways to defeat the kind of *media regime* operated by Berlusconi. For instance, for Grillo one of the best ways is to infiltrate it with the information that they are trying to censor (Grillo,

2004: 405). Grillo in fact uses the Web to "perforate" the system, to make the public aware of different truths (Grillo, 2004: 405).

Origins, Features, and Numbers of an Italian Blog: Beppegrillo.It

Beppe Grillo's blog is a compelling example of the political challenge posed by new media to older mainstream media. In a country like Italy, where politicians control both media and government, beppegrillo.it can be seen as an archetype, a model of a new type of civic engagement that has the potential to reform Italian politics.

In the words of Grillo, a "blog is an amazing thing that connects people," virtually and practically. Beppegrillo.it aims at providing a free platform for all citizens who are willing to communicate and share information, regardless of their political views (Grillo, n.d. -a). Facilitated by a direct link with the online social networking portal Meetup.com,[5] the blog aims at being the first point of call for people who are looking to engage both online and off in a fight against the monopoly grip on truth exercised by politically biased media.

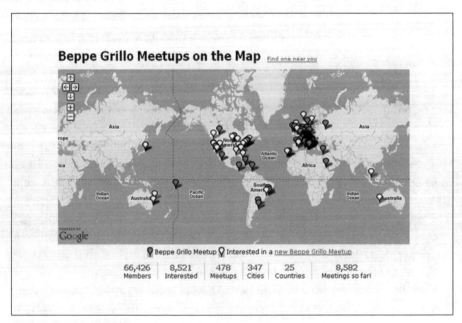

Figure 9.1. Beppe Grillo Meetup groups on the map.

To date, the Meetup.com group category "Friends of Beppe Grillo" has around 66,000 members, themselves organized in 478 groups located in 347 cities in 25 different countries (Figure 9.1).[6] The friendship groups meet regularly; they have organized more than 8,000 meetings, and sometimes, when possible, Grillo himself attends their meetings, either in person or in video-chat.[7] Moreover, this multitude of Meetup groups is slowly shaping up into a self-aware international committed network of political activists capable of organizing itself beyond geographical boundaries, independently from the blog. The network uses Meetup to coordinate itself (as in the case the V-day protest—see more below) and harness the power of free software such as Skype, the free Internet-based phone software, to organize International online meetings between its members and discuss the future course of actions.[8]

The Blog's Features: *La Settimana*

Less than a year after its first post, in mid-December 2005, beppegrillo.it was voted *best Internet site* in the category "News and Information" for the renowned *WWW 2005 Prize*. The yearly prize, organized by *Il Sole 24 Ore*, the most popular Italian daily financial newspaper, was awarded to Grillo's blog "for the interactivity with the public, the ample documentation on the Internet and the commitment to tackle topics of use to citizens" (Grillo, 2005b). Like any standard blog, beppegrillo.it stores posts by relying on two different kinds of archives: one is organized by months, while the other is organized according to categories, of which there are ten (see note 13). The blog, however, displays some unique features that are strong indicators of the political character inherent to the blog. One of the most important features of Grillo's blog is the way it seeks to widen its own reach through *La Settimana* ("The Week"; see Figure 9.2).

La Settimana is a print magazine containing the posts published on the blog during the previous week. Recently, Grillo has also begun to post a video version of *La Settimana* on the free-video-hosting platform YouTube.com (See StaffGrillo, 2006a, 2006b, 2006c), widening even more the reach and appeal of the magazine. The videos of *La Settimana* are viewed by hundreds of thousands of people each week. In Italy, where 52 percent[9] of the population is disconnected from the Web, this rather old-fashioned weekly pamphlet is an attempt to export information from the Web onto the streets. In an editorial published in the first issue of *La Settimana*, Grillo called for taking what he called a step back. He wrote (making fun of Lenin) that *La Settimana* was in effect "one step back in order to go forward" (Grillo, 2006a). What he meant was that the blog uses a traditional method of distributing political information (the printed pamphlet) in order to bridge two different worlds: the world of bits with the world of bricks.[10]

Volume 1, numero 01

La Settimana

di Beppe Grillo

Temi dal blog www.beppegrillo.it

Editoriale

"La Settimana" è un ritorno all'antico, al volantinaggio, alle copisterie nelle cantine. Una nuova Carboneria. Stampatelo e diffondetelo, ma senza dare nell'occhio, come se faceste parte di una P3, una sorta di P2 buona. "La Settimana" è un passo indietro per poter andare avanti: informazione nata in Rete e portata sulla strada. Un oggetto di modernariato mediatico. Usatelo! In dosi massicce è utile per controllare i nostri dipendenti e riportarli al loro unico ruolo: quello di amministratori della cosa pubblica.

Beppe Grillo

Dipendenti co.co.co.
Politica
02.01.2006

Siamo tutti preda di un incantesimo. Un incantesimo creato da noi. Ci siamo autoipnotizzati! Abbiamo creato un gruppo di persone che parla, che fa e disfa, che influenza le nostre vite. Giornalisti che sentenziano (in base a quale competenza?), politici (ma ormai che mestiere è?), finanzieri che creano soldi (ma i soldi non si creano), ministri (senza conoscenza di ciò che gestiscono), dirigenti di azienda (che pensano di essere loro i padroni al posto degli azionisti). Signori del nulla. Un incantesimo malato, che premia i peggiori, quelli che non creano valore, quelli che non hanno una professione. E che fanno della politica e dell'informazione una cosa loro, privata, non un servizio. Un incantesimo che emargina chi vuole cambiare, che fa emigrare i nostri migliori ragazzi, che ha impoverito la nazione. Esorcizziamoli, proviamo ad annullare l'incantesimo, questa gente non ci serve, è indispensabile solo a sé stessa. L'incantesimo si può spezzare con qualche amuleto. Incominciamo con un primo amuleto, dedicato ai nostri dipendenti: una proposta di legge popolare per ridurre a due sole legislature la possibilità di essere eletto al parlamento, italiano o europeo. Ovviamente con effetto retroattivo. Basta con i pomiciniandreottimastellacasinidalemavi olanterutelli. No ai politici a vita, si ai dipendenti a tempo determinato.

H5N1: informazione diretta
Salute/Medicina
03.01.2006

La notizia più "submarine" di questo inizio 2006 è l'influenza aviaria (H5N1). Non ne parla più nessuno. Gli italiani che hanno comprato decine di migliaia di dosi di tamiflu cominceranno a pensare che i temuti effetti dell'influenza aviaria (milioni di morti nel mondo) siano in realtà una brillante idea marketing delle società farmaceutiche. Il tamiflu non si trova più da nessuna parte (esaurito) e neppure notizie sull'epidemia. Che sia un caso? Sul tamiflu non posso aiutarvi, ma sulla diffusione dell'influenza aviaria qualche informazione posso darvela. L'ultima morte sospetta per H5N1 è avvenuta lunedì 2 gennaio 2006 all'ospedale Sulianti Saroso di Jakarta in Indonesia. La persona colpita dalla malattia era spesso in contatto con pollame d'allevamento ed è il dodicesimo decesso di H5N1 in Indonesia. La contabilità mortuaria è a questo punto di 75 morti per H5N1 dal dicembre 2003, tutti in Asia: -Cambogia 4 -Cina 3 -Indonesia 12 -Tailandia 14 -Vietnam 42 Ad oggi il pericolo di pandemia dai flussi migratori degli uccelli è considerato marginale. L'influenza si diffonde con l'aereo e il pollame contaminato. Queste informazioni si trovano in un sito che ogni giorno fornisce tutte le informazioni sull'H5N1: www.promedmail.org, usatelo per saperne di più e iscrivetevi alla sua newsletter giornaliera. E, se la notizia che ricevete è importante, inviatela con una mail ai giornali, così, tanto per informarli, e in copia anche a Storace.

Figure 9.2. *La Settimana*, January 9, 2006 (Grillo, 2006a). Permission courtesy of www.beppegrillo.it.

One Year In Numbers, a First Assessment: Hints on the Future Trajectory of a Rising Star?

One of the most common ways to measure the "authority" or importance of a blog is to count the number of other bloggers that link to it. This is the methodology used by Technorati.com, the leading blog search engine to determine the search rank of blogs. According to Technorati, beppegrillo.it ranks 9 in a list currently tracking and ranking over 112 million blogs.[11] There are 13,087 other blogs that link to it.[12] According to a 2006 Pew Internet & American Life Project Report, in the American Blogosphere the average number of inbound links to any blog is about 13 (Lenhart & Fox, 2006: p. v).

Calculating the number of inbound links, however, is only one way to evaluate the importance of a blog. This chapter argues that alongside Technorati's methodology, it is also relevant to assess and measure (1) the quantity and quality of the comments posted on a blog by its daily visitors; (2) the by-products deriving from the ongoing activity of the blog.

> Data drawn on comments (i.e., the chosen topics, the length and content quality of the text posted) give a clearer picture of the cultural and political spectrum of the community orbiting around a blog. In fact comments are the prime means by which the readers can actively enter the conversation, engage, and influence the discussions about the issues raised on the blog.
>
> Moreover, the assessment and evaluation of the growing influence of a virtual political space such as beppegrillo.it cannot only be related to its online and textual existence, that is confined within the electronic boundaries of its Internet domain. A thorough assessment of it needs to look beyond those borders and take into consideration the wider footprint of such endeavor, that is to say understanding and monitoring the effectiveness and influence of its many by-products. In the specific case analyzed in this chapter, it means to take into account by-products such as Meetup groups, campaigns, *Liste Civiche* (Civic Lists, with no connection to political parties) supported and sponsored by the blog but organized independently by citizens for the 2008 General Election.

At the core of this chapter are data on beppegrillo.it drawn from its first year of life, the 12-month period starting from May 1, 2005. The period is relevant because it culminated in the Italian general election held the following year, in April 2006. Data show a constant growth in number of comments, with a focus on politics.

During this overall period, 401 posts were published on Grillo's blog, in other words, a daily average of 1.18 posts. Each post received an average of 1,154.92 comments (see Figure 9.3).

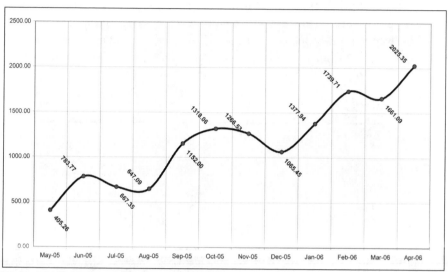

Figure 9.3. Average comments per post per month.

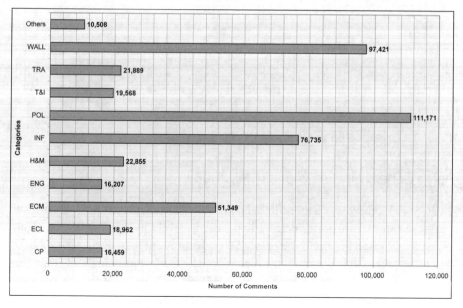

Figure 9.4. Number of comments per section.[13]

The most active site of comments was "Politics," which received more than 111,000 comments for the whole year—almost a quarter of the total number of comments posted on the blog (463,000). On average, the subject of politics scored over 1,300 comments per post. The second in the list is the general category named the Wailing Wall, 21.04 percent, but with the highest average of comments per post, 1,476.08 (see Figure 9.4 and see note 13 for the list of categories).

The overall number of comments (see Figure 9.5) grew in one year by 368.87 percent. It jumped from 17,021 comments (May 2005), to 62,786 (April 2006).

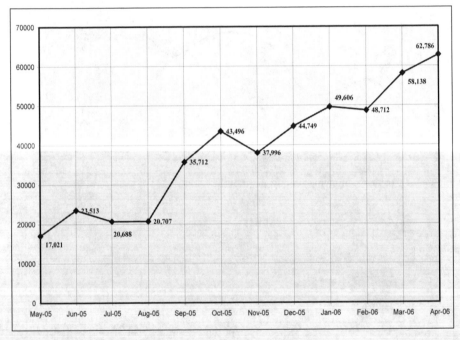

Figure 9.5. Comments per month—one year-period.

Italy's General Election (April 9–10, 2006)

In May 2005, there was a monthly average of 405 comments per post. During the year, the number of posts increased constantly, while in April 2006 the figure topped 2,025 (see Figure 9.3)—nearly 500 percent more than during May 2005.

Given the political nature of beppegrillo.it, it is not surprising that it became increasingly active around the 2006 general election. During that month, the average number of comments per post peaked at 2,025. The message posted by Grillo the day after the election, April 11, "*C'è chi*" ("There are those . . ."), produced 4,198 comments, the highest number of comments for the whole year examined here. Grillo's message commented on the close-call victory of the center-left coalition, led by former EU Commission President Romano Prodi over the center-right coalition led by Silvio Berlusconi. The closing words of the message can be interpreted as the motto of the blog and its community: "There are those who looked up at the ceiling from under the covers [of their bed] and decided never to give in" (Grillo, 2006b). However looking closer at the content of the comments the impression that one gathers is of a jubilant optimism mixed with several degrees of caution; there is a sense of shared faith in changing and improving the quality of life of the Italian people, and at the same time an acknowledgment that the close-call election victory shows a divided country and makes things more difficult for Prodi's coalition. Jubilant manifestations of hope coexist with skeptic fear that a difficult path lays ahead for the new government, that hard work is needed to heal Italy. The first comment unsurprisingly is: "*evviva è finita!!! evviva la legalità!!*" ("Hurrah! It is over!!! Hurrah for legality!!") Then later, more cautiously, Roberto Rondini writes: "Now let's be careful of the dirt deals and above all let's start working to return straight away free information to the (many) citizens who still know nothing [. . .] 9 Million Italians [. . .] voted again, in 2006, a person like Berlusconi [. . .] How many would they be if they could listen to news? I don't mean partisan news, but simply news"[14] (Grillo, 2006b). And it goes without saying that, even though the majority of comments belong to readers who voted for Prodi, there are also comments posted by Berlusconi's supporters. Some of these, naturally, are denigratory of the thin victory of the Center-Left; others instead are particularly balanced and democratic in their approach to the debate. For instance, Beppe Boselli asks for respect for those 49 percent of Italians like himself who voted for a center-right coalition and he wishes best luck to Prodi's coalition, hoping that Italy will be governed better than in the past (ibid.).

Beppegrillo.It: Campaigns

In its first year beppegrillo.it organized a number of grassroots campaigns, ranging from efforts to protect and sustain scientific research to economic and political issues. It often took a firm stand on matters that have been underrepresented or misrepresented within the mainstream media. Of these campaigns, two stood out for their success in engaging the public participation and the interest that surrounded them: *Parlamento Pulito* (Clean up the Parliament) and *Le Primarie dei Cittadini* (Citizen Primaries). These two campaigns and their organizing processes represent an important blueprint of beppegrillo.it: on the one hand they show the strengths of the blog in functioning as a virtual agora where its community can actively debate and engage with social and political matters that are often neglected by mainstream media; on the other hand they raise questions about the organizational process of the campaigns, the strength of the involvement of the citizens, the procedures of accountability inherent to these two campaigns, and the ultimate political impact of the campaigns.

Clean up the Parliament

Clean up the Parliament stemmed from an initiative of the Beppegrillo Meetup group in Milan and aimed to inform the Italian public of a simple but rarely discussed fact: that within the Italian parliament there are several *Deputati* and *Senatori*[15] who, although they have been convicted by the courts, are still allowed to represent their constituents (see Gomez & Travaglio, 2006). The ultimate aim of the initiative was to protest against the lack of adequate legislation for preventing such corruption (Grillo, n.d.-b). Beppe Grillo and his fellow bloggers published 25 posts which received 29,382 comments (on average, 1175.28 per post) and raised almost 60,000 euros, most of which was used to purchase a one-page advertisement in the *International Herald Tribune* (*IHT*) that drew attention to the problem and asked the members of parliament who were among those convicted to resign (Grillo, 2005a, 2005c). The campaign raised some interest and support overseas, most notably from Anupam Mishra, secretary of the Gandhi Peace Foundation of New Delhi in India, who, in a long letter addressed to Grillo and then posted on the blog, commented: "We congratulate you on such a courageous advert and the important piece in the services of civil society. [. . .] We have circulated [widely] your inspiring advert to some news channels and Hindi newspapers [. . .]" (Grillo, 2006f).

Citizen Primaries

On the 16th of October 2005 for the first time in Italy, a political coalition, *L'Ulivo* (The Olive Tree) asks its supporters to vote for one from a list of seven candidates as the 2006 general election prime minister nominee. These primaries had an unexpectedly large turnout of over 4 million voters, with 74 percent of the votes going to Romano Prodi, who eventually defeated Berlusconi's coalition in the General Election.[16]

At the beginning of 2006, in preparation for the election, the center-left coalition presented to the press and their electorate a first draft of their political program (Unione, 2006). The 281-page document was judged by many commentators as too long and complicated to be able to reach out to the electorate (Triglia, 2006).

Echoing the successful 16 of October initiative, in an attempt to overcome the 281 pages of flawed text, in January 2006, Beppe Grillo launched a new campaign entitled *Citizen Primaries*. The campaigners used beppegrillo.it as a platform from where to stir up a political debate among politicians and citizens on topics that according to Grillo, his staff, and his readers, should be at the core of the political program of Prodi's coalition. Its goal was to produce a new program (from below) of political and social reforms that would reflect more adequately the people's needs.

In the first post outlining the new campaign Grillo wrote: "Up till now it's been our employees[17] who've done Primaries. Now is the time for the employers to do Primaries. From today [8 January 2006] I'll be publishing proposals about important issues like Energy, Transport, Electoral Regulations and I'll be assisted by recognised experts in the field, so that I can receive your comments" (Grillo, 2006c). The message continues with an invitation to the politicians to take part in the debate.

Under the category "Citizen Primaries" there were 11 Messages divided in 4 subsections: Energy (*energia*), Health (*salute*), Information (*informazione*), and Economics (*economia*). The first post is dated January 8, 2006 and the last one April 4 of the same year, six days before the general election. Overall the messages posted on this section of the blog received 16,458 comments, on average 1,496.18 comments per post.

The proposal on energy (Grillo, 2006c) was the most commented up on throughout the whole campaign (7,846 comments[18] or 47.67 percent of the total number of comments; health received 3,306 messages, information 2,894, economy 2,412). It focused on the optimization and reduction of energy waste ("half of the energy consumption is wasted" (Grillo, 2006c) and on using new technologies to allow the development of renewable sources, such as wind power and solar energy. Among the many suggestions proposed by Grillo and his community were (1) to increase the incentives for reducing energy waste and for the production and the use of alternative energy-powered vehicles, (2) to use public roofs for solar panels, (3) to stop the production of nuclear energy, (4) and to fight against the monopoly of oil companies

As the campaign took place in the four months before the general election, it is not surprising to find out that not all comments were directly linked to the topic of the posts, some in fact were more election driven and addressed topics such as the importance of voting and the merits and shortcomings of various candidates. Comments were posted by people from all walks of life and political backgrounds, and they critiqued both parties for their short-sighted energy policies (Grillo, 2006c).

A few months later on June 8, 2006, Grillo on behalf of the community that developed around his blog met the newly appointed Prime Minister Romano Prodi and gave him a book containing the suggestions for the government collected through the blog during the primaries (Grillo, 2006d; StaffGrillo, 2006d), and an "end of employment" letter warning him that he would be dismissed from his post if he decided not to give the appropriate attention to the proposals that resulted from his blog's Citizen Primaries. The letter was singed "the Employer" (Grillo, 2006e). The meeting was video-recorded and posted on Google Video and YouTube.com. Many users uploaded copies of the video. Since the first posting (June 15, 2006) many hundred thousands of people have watched the video online; for instance, to date just one of the many versions available on the two hosting platforms has been viewed 73,096 times (Retrieved on April 17, 2007; see StaffGrillo, 2006d).

Accountability, Representation, and the Problem of Lack of Transparency

Although these two campaigns used innovative techniques to facilitate grassroot political engagement, some elements of their organization and strategies were not entirely democratic or transparent.

Clean up the Parliament was an initiative clearly originating from below and outside Beppe Grillo's entourage. It represented a group of people who were eager to spark a debate on an important issue related to the legal and moral status of the members of the Italian Parliament. It raised money to achieve a very simple goal: to raise awareness on the issue through publishing an appeal in a national newspaper. The page appeared however in the *International Herald Tribune*, because, so Grillo explained (Grillo, 2006e), the Italian newspapers he had contacted refused to publish the appeal. However, although many praised the initiative, some of the blog's readers openly disagreed with Grillo and his way of conducting the campaign: they criticized the use of his own name as beneficiary of the donations, instead of opening a bank account with the name of the initiative, as some suggested, or simply, as others wrote, use an online payment service, such as Paypal, which would have been easily verifiable. After all most of the donors were people using the Internet and probably familiar with online payments. Criticism also was on the lack of full transparency in the using of the donations: Grillo in fact never published the certified copy of the invoice of the almost 60.000 euros payment made to the *IHT*. He also wrote that he intended to publish the initials and the amount donated by each contributor to cover the costs of the initiatives (Grillo, 2006e), but neither the list nor the invoice was ever uploaded on the blog. Moreover, readers pointed out the undemocratic flair of Grillo's behavior: without previous discussion with the donors through the blog, he autonomously decided to go ahead and publish a page in the *IHT* only with the description of the campaign but not the names of the politicians convicted—as it was in the original proposal from which the whole campaign had stemmed.[19]

The other campaign, the Citizen Primaries did not originate from an open debate about which topics and issues should be addressed. In fact, it was never clear how Grillo came up with the original proposals on the four topics chosen for the campaign and who exactly his "experts on the field" were (see Grillo, 2006c). Apparently, Grillo was the *Primo Mobile* of the proposal and only afterward the readers of the blog were able to post comments regarding those proposals and suggest new ones. Readers however were never requested or put in a position to cast their vote. In fact, notwithstanding Grillo's bold statement that the proposals were "a concrete example of direct democracy and participation of people in public affairs" (Grillo, 2006d), there was no true and clear polling mechanism provided to the users to reach an agreement on issues and solutions. There was no open mechanism available to readers to propose topics and discuss them. A dedicated Web-forum for the campaign for instance could have helped make the whole debate more transparent and attach to the Citizens Primaries a sense of true belonging for the readers of the blog. Moreover, an Internet-based survey could have clarified the preferences—that is, the numbers—of the public. Because of this lack of transparency and plain democratic mechanisms, it is even less clear how those original proposals changed in accordance with the inputs received from the sixteen thousand comments posted on the blog throughout the entire length of the campaign. The behind-the-scene remains hidden to the public, in fact it is not clear how many emails, proposals, and ideas Grillo received before and after posting those messages and how he used them. Moreover, primary and secondary sources in the blog are rarely ever given.

For more than a year since the end of the primaries, the final electronic version of the printed hard-bound document which Grillo hand-delivered to Prime Minister Prodi was absent from the blog. To some this is evidence of Grillo's flawed approach to the problem of accountability and transparency in the process of information sharing. He films himself bringing the document to Romano Prodi, saying "I bring this on behalf of my bloggers"(Grillo, 2006d), but he fails to provide the electronic version of that document to his readers, an act of transparency that would allow the beppegrillo.it community to verify and analyze the document. On this matter Alessandro Esposito commented: "Dear Mr. Grillo, I am 17 years old and I have been following you since it has been again possible to get in touch with you via your blog. I would like to receive a copy [. . .] of the document you delivered to our employee Prodi." (Esposito, 2006)

Conclusions: Lights and Shadows

Those who read and comment on Grillo's posts are members of an active public inspired by the comedian. In addition to posting thousands of comments on the blog, they post videos on external platforms, create and participate in social and political campaigns, publicize the blog and the work of its community, and organize regional and international gatherings via Meetup. In these ways they fight against the political establishment and actively attempt to give life, substance, and direction to a form of politics that aims to create a more democratic alternative to that status quo. This is a public that understands the value of the democratic political process and the importance of new communication media to be successful in that process.

Notwithstanding its evident success, Grillo's blog is not without shadows or immune to criticisms. Enrico Lincetto, one of the thousands of readers of the blog, in a comment he posted in response to Grillo's visit of Prodi (June 8, 2006), questions the comedian's self-proclaimed status as *messenger of the people*. The document delivered to Prodi is—in Grillo's view—a document allegedly listing the problems of the Italian people and therefore it "should be taken in great consideration by [. . .] the Government." So, addressing the subject rather frankly, Mr. Lincetto asks: "Are the majority of the Italian people participating in Beppe Grillo's blog? How can you interpret the will of the Italian people when you represent no one?" (Lincetto, 2006).

The question is an important one, but the answer is not as simple. On the one hand Beppe Grillo does not officially represent anyone because he has not been elected. On the other hand the growing number of people that turn to his blog to read and assiduously comment on issues that he raises seem to embrace Grillo as a de facto leader. Commenting on his unusual position as a political guru, Grillo says, "People continuously write to me on my blog to tell me that I am the only person who can say certain things [. . .] but really I am only a comedian, I shouldn't have this weight" (Povoledo, 2006).

The latest episode in the Clean up the Parliament initiative is perhaps the most convincing evidence of his political clout. Almost two years after the appeal published on the *IHT*, on September 8, 2007, Grillo and his followers organized a protest called the *V-Day* or *Vaffanculo Day* (*Vaffanculo* means "fuck off" in Italian). On the day commemorating the Italian armistice in World War II (September 8, 1943), Grillo asked his public to gather in the squares of their cities throughout Italy and the world and to sign a petition to propose a new law to the Parliament. The proposed law is composed of three different elements: (1) candidates convicted by courts of law should be forbidden from running for public office; (2) political careers should be limited to only two terms; and (3) that the members of the Parliament should be directly chosen by the people (Grillo, 2007a).

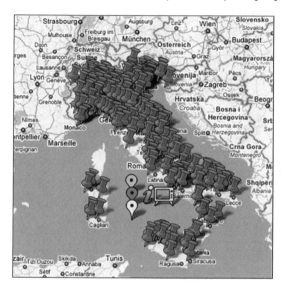

Figure 9.6. Map of the location of the petition desks for V-day throughout Italy. Source: Google Maps.[20]

The V-day was a success both in terms of numbers and media exposure: 350 thousand people gathered in more than 200 cities worldwide (see Figure 9.6 for ma of V-day activity throughout Italy). The meetings were organized through the blog and through Meetup.com. In the aftermath of the event, the issue was debated in the pages of the Italian newspapers and on television,[21] and it sparked harsh reactions from politicians from both sides of the Parliament. Commenting on this success, Grillo said: "I was really surprised. I didn't expect such a big turnout. . . . What happened out there was the release of a virus that's about to attack the political class. But in this case there's no vaccine" (Povoledo, 2007).

Riding the momentum of V-day, soon after the event ended, Grillo launched Liste Civiches (civic lists), a new challenge to the political establishment. Liste Civiches are collaboratively created lists of local administrators who meet the standards of the Grillo community. For the lists to get the Beppe Grillo stamp of approval they must fulfill a set of requirements; for instance, they cannot be linked to political parties; their members must have a clean record, each candidate should reside in the same location as his or her constituents, and candidates may not have served previously more than one term in office—either at local or national level (see Grillo, 2007b). Grillo, however, is not attempting to create a new political coalition, but rather he stresses that each of these lists should be autonomous. He says, "I am not promoting any Civic List, neither local, nor national. The participants of the V-day do not lend their voices to anyone. They are megaphones of themselves. They are citizens that do their own politics" (*La Repubblica*, 2007a).[22]

Thus, willingly or not, the success of his blog and of the initiative like the V-Day, coupled with daring and innovative projects such as the *Liste Civiche*, show that Grillo is the (nonelected) representative of the dissatisfied public orbiting around his blog. Not surprisingly, many of the comments posted on the blog ask the comedian to enter politics in a conventional way via the election process (see for instance comments in Grillo, 2006c). Seen from this perspective, beppegrillo.it represents a new emerging trend in the political sphere: a politics outside formal politics, and according to Grillo, this trend is *a blueprint for the future* (Povoledo, 2007).

To give this future stable foundations and more credibility, however, Grillo (and others in his position) needs to confront the moral obligations that bear upon the shoulders of any political representative within any democratic environment: he needs to be accountable and his actions need to be fully transparent, or he will not be able to escape the accusations of mere populism that many of his critics attach to his endeavors.

Notes

1. In 2005, *Time Magazine* included the *rebel* Grillo in the list of the 37 European Heroes of 2005 "who illuminate and inspire, persevere and provoke" (Israely, 2005).
2. The two networks' groups combined account (on average) for 87 percent of the daily share of the Italian television audience (*La Repubblica*, 2006a).
3. No other democratic country worldwide has witnessed similar ties between political leadership and media ownership (Gomez & Travaglio, 2004: p. xvii).
4. Berlusconi, was accused, among other things of having ties to the Mafia (see Gomez & Travaglio, 2004: 28–62).
5. In the center right-hand-side of the webpage there is a red bordered logo labeled Meetup. Underneath the logo there are two sets of lists: the first is a list of Italian cities; the second is a list of international cities. Meetup.com is an online portal that facilitates social networking. The stated aim of the portal is to help "people find others who share their interest or cause, and form lasting, influential, local community groups that regularly meet face-to-face" (Meetup, n.d.). Searching by topics and/or postal code, Meetup members can find and then join other people organized in groups that meet regularly. To create a group, the organizer is required to pay a fee of 19 dollars.
6. http://beppegrillo.meetup.com/about/?gmap=1, Retrieved February 15, 2008. The data relative to Meetup.com have almost doubled since May 1, 2007; back then "Friends of Beppe Grillo" had around 33,000 members, themselves organized in 199 groups located in 157 cities in 15 different countries.
7. See for instance the International Meetup (January 19/20, 2008), organized by the Beppe Grillo's Friends Amsterdam Meetup Group; during that meeting Beppe Grillo himself con-

nected to the audience via video through Skype. A programme of the meeting is posted here: http://beppegrillo.meetup.com/434/messages/boards/view/viewthread?thread=3871169. A recording of his message is uploaded on the Ustream TV: http://www.ustream.tv/recorded/ GraU4QNDsf,HhpRDmDhnscjXC2JHyRUF (Retrieved February 2008)

8. Personal communication with Ethel Chiodelli, Organizer of the London Beppe Grillo Meetup Group.

9. Source: the Internet World Stats Website, March 31, 2006 (http://www.internet worldstats.com/stats9.htm).

10. Sifting through the many thousands of comments posted on beppegrillo.it, I found one that indirectly seems to be linked, although it was never acknowledged by Grillo, with the origins of *La Settimana*: just over a month before the publication of the first issue of *La Settimana*, Vincenzo Curcio wrote a comment on one of Grillo's posts (Grillo, 2005c, the comment is dated: November 29, 2005 15: 39) addressing the issue of how to make available the content of the blog to those who cannot use the Internet. Mr. Curcio suggested that on a weekly basis the various arguments discussed on the blog can be collected in a few pages and be published in those newspapers—such as *City, Metro, Leggo*—that are distributed freely in many Italian cities (in Grillo, 2005c).

11. To date Technorati is "currently tracking 112.8 million blogs and over 250 million pieces of tagged social media" (http://www.technorati.com/about/, data retrieved February 21, 2008).

12. Data retrieved from http://www.technorati.com/pop/blogs/ February 21, 2008. It is important to note that this is a number that has doubled in less than a year: in April 2007 beppe-grillo.it was number 18 in the list and it had 6,974 inbound links (Date: April 13, 2007).

13. The ten categories are: Primarie dei Cittadini (in English CP, Citizen Primaries), Ecologia (ECL, Ecology), Economia (ECM, Economics), Energia (ENG, Energy), Salute/Medicina (H&M, Health/Medicine), Informazione (INF, Information), Politica (POL, Politics), Tecnologia/Rete (T&I, Technology/Internet), Trasporti/Viabilità (TRA, Transport/Getting About), Muro del Pianto (WALL, Wailing Wall).

14. The original is in Italian; translation is mine: *"Adesso occhio agli inciuci e soprattutto al lavoro per restituire subito un'informazione libera ai cittadini (tanti) che ancora non sanno nulla o quasi. 9 milioni di italiani—dicesi 9—votano ancora, nel 2006, uno come Berlusconi; non è che magari è perchè. . . . te Beppe che ne pensi ? Quanti sarebbero se potessero ascoltare delle notizie? Non di parte, solo delle notizie."* Roberto Rondina, November 4, 2006 16: 32.

15. The Italian Parliament is formed by two chambers: *Deputati* (deputies) are those elected to represent the people in the *Camera dei Deputati*—the Lower House—and *Senatori* are those elected for the Senate, the Upper House.

16. http://www.perlulivo.it/2005-primarie/primaria-2005.html

17. Since he started the blog, the ideas of "politicians as employees" and "citizens as employers" have been two of Grillo's more successful slogans, and the Citizen Primaries campaign played along those lines to remark the temporary nature of the political mandate of those who are elected, something that seems forgotten in a country like Italy, where there is very little turnover among the members of the political class.

18. Comments are in Italian and are available at: http://www.beppegrillo.it/2006/01/ primarie_dei_ci_3.html

19. See for instance the comments of Antonio D'Arpa—November 22, 2005 16: 54—and Silvia Landi—November 22, 2005 17: 37—in Grillo, 2005c.

20. Il V-Day nelle piazze d'Italia, 8 settembre 2007, Google Maps: http://maps.google.it/
maps/ms?ie=UTF8&hl=it&t=h&num=200&start=0&om=1&msa=0&msid=105899767908
383675040.00043790d3bb18fa8203b&ll=42.114524,11.25&spn=10.053068,20.566406&z=
6 (Retrieved February 21, 2008).

21. For articles and news about the V-Day, see the section of beppegrillo.it dedicated to the event;
it is available at: http://www2.beppegrillo.it/vaffanculoday/

22. From *La Repubblica*, 2007a. The original text is in Italian; translation is mine. "Io—puntu-
alizza—non parteciperò a nessuna manifestazione nei prossimi mesi. Non sto promuoven-
do la presentazione di nessuna lista civica, nè locale, nè nazionale. La loro voce, i partecipanti
del V-day non la prestano a nessuno. Sono i megafoni di sè stessi. I cittadini che si fanno
politica."

References

Arie, S. (2003, July 5). Italian TV accused of censoring furore over Berlusconi jibe. *The Guardian*.
Retrieved April 1, 2007, from http://www.guardian.co.uk/international/story/
0,3604,991967,00.html

BBC (2003, July 10). Press upset at Italy–Germany row. *BBC*. Retrieved April 1, 2007, from
http://news.bbc.co.uk/1/hi/world/europe/3055869.stm

Blatmann, S. (2003). *A media conflict of interest: Anomaly in Italy*. Retrieved April 1, 2007, from
Reporters without Borders Web site: http://www.rsf.org/article.php3?id_article=6393

Esposito, A. (2006). "Comment," posted June 9, 2006 at 12:33, in Beppe Grillo, 'Il dipendente
Prodi riceve i risultati delle Primarie dei Cittadini' in *Il Blog di Beppe Grillo*, posted June 8,
2006. Retrieved April 1, 2007, from http://www.beppegrillo.it/2006/06/il_dipendente
_p_1.html. The translation of the comment from the Italian is mine.

Glaser, Mark (2005). "What Really Makes a Blog Shine" in Handbook for Bloggers and Cyber-
Dissidents published by Reporters without Borders. Retrieved September 2006, from
http://www.rsf.org/rubrique.php3?id_rubrique=273

Forbes Magazine (2007, March 8). The world's billionaires. *Forbes.com*. Retrieved April 1, 2007,
from http://www.forbes.com/lists/2007/10/07billionaires_Silvio-Berlusconi-family_EEPT.
html

Freedom House (2006). *Freedom of the press 2006—A global survey of media independence*. Retrieved
April 1, 2007, from Freedom House Web site: http://www.freedomhouse.org/
template.cfm?page=271&year=2006

Ginsborg, P. (2003, May/June). The patrimonial ambitions of Silvio B. *New Left Review, 21*,
21–64.

Gomez, P., & Travaglio, M. (2004). *Regime*. Milano: BUR Futuropassato—Rizzoli Editore.
(Translation from Italian is mine.)

Gomez, P., & Travaglio, M. (2006). *Onorevoli wanted*. Roma: Editori Riuniti.

Grasso, A. (1992). *Storia della televisione italiana*. Milano: Garzanti. (Translation from Italian is
mine.)

Grillo, B. (n.d.-a.). Help. *Il Blog di Beppe Grillo*. Retrieved April 1, 2007, from
http://www.beppegrillo.it/eng/aiuto.php

Grillo, B. (n.d.-b). Enough! Clean up parliament! *Il Blog di Beppe Grillo*. Retrieved April 1, 2007, from http://www.beppegrillo.it/eng/condannati_parlamento.php

Grillo, B. (2004). 'Postfazione', in Gomez, P., & Travaglio, M. (2004). *Regime*. Milano: BUR Futuropassato—Rizzoli Editore. (Translation from Italian is mine.)

Grillo, B. (2005a, November 22). Clean up parliament, appeal from the blog www.beppegrillo.it. *International Herald Tribune*, Retrieved April 1, 2007, from http://www.beppegrillo.it/immagini/beppe_ht.pdf

Grillo, B. (2005b, December 16). The voice of the Susa Valley/9. *Il Blog di Beppe Grillo*. Retrieved April 1, 2007, from http://www.beppegrillo.it/eng/2005/12/the_voice_of_the_susa_valley_9.html

Grillo, B. (2005c, November 22). *Stand up! Clean up! The parliament . . .* Retrieved October 28, 2007, from http://www.beppegrillo.it/2005/11/stand_up_clean.html

Grillo, B. (2006a, January 9). Editoriale. *La Settimana, 1*, 01. Retrieved April 1, 2007, from http://www.beppegrillo.it/magazine/archivio/lasettimana2006–01–09.pdf

Grillo, B. (2006b, April 11). C'e' chi . . . *Il Blog di Beppe Grillo*. Retrieved April 1, 2007, from http://www.beppegrillo.it/2006/04/ce_chi.html (Translation is from the English version: http://www.beppegrillo.it/eng/2006/04/there_are_those_1.html)

Grillo, B. (2006c, January 9). Citizen primaries: Energy. *Il Blog di Beppe Grillo*. Retrieved April 1, 2007, from http://www.beppegrillo.it/eng/2006/01/citizen_primaries_energy.html

Grillo, B. (2006d, June 8). Il dipendente Prodi riceve i risultati delle Primarie dei Cittadini. *Il Blog di Beppe Grillo*. Retrieved April 1, 2007, from http://www.beppegrillo.it/2006/06/il_dipendente_p_1.html, the English translation at: http://www.beppegrillo.it/eng/2006/06/our_employee_prodi_receives_th_1.html (Translation from the blog)

Grillo, B. (2006e, June 8). Hand-delivered registered letter. *Il Blog di Beppe Grillo*. Retrieved April 1, 2007, from http://www.beppegrillo.it/eng/immagini/Raccomandata%20ING%20a%20mano%20per%20Prodi.pdf

Grillo, B. (2006f, January 22). Cleaning up the Indian parliament. *Il Blog di Beppe Grillo*. Retrieved April 1, 2007, from http://www.beppegrillo.it/eng/2006/01/cleaning_up_the_indian_parliam.html

Grillo, B. (2007a). V-day, 8 settembre nelle piazza italiane. *Il Blog di Beppe Grillo*. Retrieved October 30, 2007, from http://www2.beppegrillo.it/vaffanculoday/

Grillo, B. (2007b, October 10). Liste civiche/1. *Il Blog di Beppe Grillo*. Retrieved February 21, 2008, from http://www.beppegrillo.it/2007/10/liste_civiche1.html

Guardian, The (2003a, July 2). MEPs' fury at Berlusconi's Nazi jibe. *The Guardian*. Retrieved April 1, 2007 http://www.guardian.co.uk/eu/story/0,7369,989630,00.html

Guardian, The (2003b, February 17). Millions worldwide rally for peace. *The Guardian*. Retrieved April 1, 2007 http://www.guardian.co.uk/antiwar/story/0,12809,897098,00.html

Held, D. (1996). *Models of democracy*. Stanford, CA: Stanford University Press.

Internazionale (n.d). *Biografia di Beppe Grillo*. Retrieved April 1, 2007 http://www.internazionale.it/beppegrillo/page.php?pagid=grillo_biografia

Israely, J. (2005, October 10). Seriously funny. *Time Europe, 166*(15). Retrieved April 1, 2007 http://www.time.com/time/europe/hero2005/grillo.html

Lenhart, A., & Fox, S. (2006, July 19). *Bloggers: A portrait of the internet's new storytellers*. Washington, DC: Pew Internet & American Life Project. Retrieved April 1, 2007, from PEW Web site: http://www.pewinternet.org/PPF/r/186/report_display.asp

Lincetto, E. (2006). "Comment," posted June 8, 2006, 14: 47, it can be found in Beppe Grillo, 'Il dipendente Prodi riceve i risultati delle Primarie dei Cittadini' in *Il Blog di Beppe Grillo* posted June 8, 2006. Retrieved April 1, 2007, from http://www.beppegrillo.it/2006/06/il_dipendente_p_1.html. The translation of the comment from the Italian is mine.

Meetup (n.d.). *Meetup at a glance* [online]. Retrieved April 1, 2007, from http://www.meetup.com/about/

Povoledo, E. (2006, April 7). Italy's politics refracted, on stage and screen. *International Herald Tribune*. Retrieved from http://www.iht.com/articles/2006/04/07/news/rome-5916888.php

Povoledo, E. (2007, September 13). Protest unnerves Italy's political elite. *International Herald Tribune*, p. 2, a slightly different version is available online at: http://www.iht.com/articles/2007/09/12/america/italy.php

Repubblica, La. (2005). Le regole per la nomina dei vertici della Rai. *La Repubblica*, maggio 17, 2005, article retrieved April 1, 2007, from http://www.repubblica.it/2005/c/sezioni/politica/cdarai/nuovleg/nuovleg.html

Repubblica, La. (2006a, January 13). Nel 2005 Rai vince su Mediaset e tra i programmi trionfa Sanremo. *La Repubblica*. Retrieved April 1, 2007, from http://www.repubblica.it/2006/a/sezioni/spettacoli_e_cultura/auditel2/auditel2/auditel2.html

Repubblica, La. (2007, September 16). V-day, Grillo lancia le sue liste civiche Chi lo merita avrà il mio bollino. *La Repubblica*. Retrieved May 23, 2007, from http://www.repubblica.it/2007/01/sezioni/politica/sondaggi-2007/fiducia-istituzioni/fiducia-istituzioni.html

StaffGrillo (2006a, December). *La Settimana N 48 di Beppe Grillo* [online]. Retrieved April 1, 2007, from Youtube Web site: http://www.youtube.com/watch?v=1u7SNSRK4zY

StaffGrillo (2006b, December 16). *La Settimana N 49 di Beppe Grillo* [online]. Retrieved April 1, 2007, from Youtube Web site: http://www.youtube.com/watch?v=SfMlWZhy_n4

StaffGrillo (2006c, December 18). *La Settimana N 50 di Beppe Grillo* [online]. Retrieved April 1, 2007, from Youtube Web site: http://www.youtube.com/watch?v=z5-WoRYUVgs

StaffGrillo (2006d, June 15). Primarie dei Cittadini: Beppe Grillo incontra il Presidente del Consiglio Romano Prodi [online]. Retrieved April 1, 2007, from Google Video Web site: http://video.google.com/videoplay?docid=-7453463360129666713

The Prime Minister and the Press (2003, August 21). Documentary, Executive Producer: Stefano Tealdi, Director: Susan Gray. Broadcast by PBS. Retrieved April 1, 2007, from http://www.pbs.org/wnet/wideangle/shows/berlusconi/index.html

Triglia, C. (2006, January 28). L'Unione delle risposte mancate. *Il Sole 24 Ore*, pp. 1, 6.

Unione (2006). *Per Il Bene dell'Italia—Programma di governo 2006–2011*. Retrieved April 1, 2007, from Unione Web site: http://www.unioneweb.it/wp-content/uploads/documents/programma_def_unione.pdf

CHAPTER TEN

Textual AND Symbolic Resistance

Re-mediating Politics Through the Blogosphere in Singapore

YASMIN IBRAHIM

Much has been written about blogs as cultural and personal tools of expression in recent years (Blanchard, 2004; Herring et al., 2005; Ibrahim, 2006; Nardi et al. 2004). Blogging as an activity has become popular due to the proliferation of specialized hosting sites and user-friendly software and has in the past entered mainstream consciousness through media events such as 9/11 and the Iraq war, where mediatized mainstream narratives were complemented with personal and communal representations in the blogosphere. Blogs enable the privatization of public events and the publicization of the personal. Their association with both individual authorship as well as communal agency, their ability to conjoin public and private spheres, and their location on a global platform as a cultural tool mediated by the context of their occurrence have challenged theorists to analyze blogs in a myriad of ways.

Whether as cultural tools of expression or as personal conduits for public ruminations, Weblogs fuse global practices with local agency. They tend to be writer-centric, have a chronology, and are often link-heavy, representing hidden networks and intertextual links that situate the text within a composite ecology of information. Blogs, as an open system of communication, can support the formation and maintenance of a social network of both writers and readers (Su et al., 2005) where technology has enabled "instant, updated and frequent communication of information such as events, personal interest, thoughts and news" (Chin & Chignell, 2006).

This chapter seeks to map the degrees of empowerment, forms of cultural dissonance, and symbolic and material resistance that the blogosphere presents in the political and social landscape of Singapore by analyzing the 2006 general election in which political blogs played a significant role. According to Tan Tarn How of the Institute of Policy Studies (IPS) in Singapore, the tiny island state has "hit the global blogging big league" (cf. Crispin, 2006), reflecting the popularity of blogs as a cultural tool among the computer-savvy population. In addition, the authorities' caution against persistently political blogs prior to the general election in 2006 captures the increasing use of information and communication technologies (ICTs) for political expression as well as the expansion of the public sphere through such technologies. The chapter locates the material and symbolic resistance of the blogosphere by analyzing the discourses that seek to regulate blogs in the country.

The relationship between new media technologies and the forms of resistance they present to physical spaces of power and ideology is a resonant strand of enquiry in new media studies. The relationship between virtual space and offline society is important in comprehending the Internet as both a signifier of culture as well as a repository of culture. Christine Hine's (2000) approach to investigating the Internet highlights this double articulation, for the Internet can be investigated both as a technology shaped by culture and equally as a site of cultural production. The technologically mediated context then becomes the canvas for mapping the types of symbolic and material struggles and resistance that can be manifested through new media spaces. From the perspective of literary and critical studies and in particular post-structuralism, the occurrence of text (in new media spaces) cannot be separated from the larger cultural and social textuality out of which it is constructed (cf. Allen, 2000: 37). As such, online discourses can only be understood through the social specificity of their occurrence.

Singapore's online regulatory environment is a well-researched phenomenon (Baber, 2002, 2003; Banerjee & Yeo, 2003; Gomez, 2002; Ho et al., 2003; Lee, 2001a, 2001b, 2002; Rodan 1998), and often the thrust of the analysis has focused on how the Internet can enhance the potential for democratization in the island. This chapter, while describing the online regulatory environment that seeks to map resistance on the Internet, tackles the notion of resistance and empowerment through the construct of intertextuality. It applies the notion of intertextuality to the hypertextual environment of the Internet to trace the politics of resistance and representation in the virtual space. One of the main premises of this chapter is that the cultural context of offline society mediates the readings of text and images that occur online (Miller & Slater, 2000).

Locating Intertextuality

Intertextuality is a term which is attributed to Kristeva in an essay in 1966 to describe the interdependence any literary text has with others which preceded it. Intertextuality may be said to have its origins in twentieth-century linguistics, particularly in the works of de Saussure (Allen, 2000). Intertextuality also emerges from the theories of those more concerned with the existence of language within specific social situations or the centrifugal forces which shape the engagement with text. Text negotiates boundaries and makes connections with other texts, fields, and genres. Intertextuality can be both centrifugal and centripetal and often signifies a quest for a movement to the center as well as the periphery to explore the sources and motives upon a text. Lyotard (1971: 75) describes the occurrence of the text as the "constitution of a thick space where the play of hiding/revealing takes place." Edging away from semiotics to a wider exploration of narrative structures, this chapter's interpretation of intertextuality focuses on the centrifugal elements of culture to situate a text instead of the centripetal (i.e., the search for the source and roots of a discourse).

Intertextuality proclaims a relationship whether discreet or overt with other texts. Texts here then imply the whole ecology of media and cultural artifacts. For contemporary theorists such as Umberto Eco (2003) texts always speak to one another. This means that the text is affected by the intertextual dialogue between them. For Eco (1986), intertextuality can be signified through various intentions. These can encompass intertextuality which is conscious and direct or work at a subconscious level of cultural intertextuality which acknowledges foundational assumptions upon which a text is based.

Bahktin's (cf. Allen, 2000: 19) notion of dialogism propounds that no utterance exists alone and emerges from a "complex history of previous works and addresses itself to, seeks for active response from, a complex institutional and social context." Bakhtin's work traced much of this dialogical tradition in novels, and it is the aim here to apply it to the Internet as a cultural artifact and in particular the digital hypertext environment of new media.

In the context of Singapore, this chapter seeks to employ intertextuality to understand the dialogism between offline and online discursive spaces to discern new forms of agency and political empowerment within a given polity. The plurality of the Internet provides new spaces for recording the civil and social history of a nation. In this sense, the Internet is an unbound textual space which is always in the process of being produced. As a cognitive space it conjoins the virtual and the physical spaces producing new streams of consciousness and dialogism which often locate the social context of these utterances.

Kristeva (1984), in commenting on the novel, states that "every text builds itself as a mosaic of quotations, every text is absorption and transformation of another text." As such the novel is a specific signifying sign system which itself is a result of various sign systems. Thus the term intertextuality denotes this transposition of one sign system into another. New media technologies make meaning not only by building a new text through absorption and transformation of other texts, but also by embedding the entirety of other texts seamlessly within the new (Everett, 2003). In applying the term intertextuality to online spaces this chapter seeks to integrate the social context of its embedding to explain the politics of resistance in textual spaces.

The Social and Political Context

The social, economic, and political context of Singapore within communication studies and political science literature has been well documented and analyzed (cf. Chan, 1976; Christie, 1997; Chua, 1995; Khong, 1995; Kuo, 1983; Low, 2001; Vasil, 2000; Worthington, 2003), and it is not the intent here to debate the nature of governance but to draw the salient features in order to analyze how the blogosphere challenges some of the norms of political discourse. The nation-space can be imagined by a community (Anderson, 1991) through cultural and media artifacts, and this communal imagination can be influenced and limited by how the nation is narrated by the voices that dominate the space (Bhabha, 1990).

The political culture of Singapore is mediated by the ideology of the dominant ruling party, the People's Action Party (PAP), which came to power in 1959 and won the most recent general election in 2006. This long and sustained political legacy has left a trajectory of monologism which situates plurality as a destabilizing element in the social and political history of Singapore. Drawing on historical events such as the Maria Hertogh riots in 1950, the 1964 riots during the prophet Mohammed's birthday, and the spillover riots from Malaysia in 1969, the PAP has built a linear narrative of plurality as a violent atavistic trait and a core justification for censorship (Ang & Nadarajan, 1996). The nation as a narrative space then sustains the monologue of the PAP conceiving it as a totalizing force which consistently blocks competing sites of contesting text and ideas. Censorship is defended on moral grounds, and the blocking of competing narratives is seen as necessary to the discourse of the nation-building exercise.

In studying the media as a site of culture in Singapore, the mainstream media record the history of the nation from the temporal dimensions of the PAP's struggle, growth, and consolidation within Singapore. The PAP, by controlling the culture industry (i.e., press, print, radio, and TV), has imposed a collective memory which is sustained not just through media discourses but through institutionaliza-tion of memory through the control of archive, memory, and culture. The PAP's con-trol of the media, and the archiving of memory and history through these controlled spaces, is important in mapping the politics of resistance in the online environment. The nation is imagined and narrated through these dominant media articulations. This "imaginary economy" is constructed not just through the PAP's control of the political economy of the media but also its overarching ideology, which imposes a synthetic and modern memory that is selective and constantly obliterates and mar-ginalizes dissent and opposition voices in narrating the politics and history of the nation. Here the control of the archive and the construction of a prosthetic mem-ory constitute a site of ideological and moral power over the citizenry (Derrida, 1995).

Cherian George's (2001) metaphor of an "air-conditioned" nation illuminates the dialectics of control and comfort in Singapore. The analogy of air-condition-ing a nation-space accords great powers to the PAP where it is seen as having the ability to closely control the temperature setting of an ecological human habitat. From a political perspective the metaphor conjoins the notion of material comfort with control where the latter is a convenient trade-off for the former. This antithet-ical duality of control and comfort constructs politics as a remit that is best left to the technocrat to maneuver and one in which the ordinary citizen is not qualified to participate. The delimiting of politics to an elite echelon of technocrats has nar-rated politics from the onset of nation-building as a domain best removed from civic participation and individual articulation. This is often augmented by the "respon-sibilization" debate where an individual's political participation is associated with responsible citizenry where alternative views are presented as destabilizing the communitarian goals of the nation. One of the commonplace constructs which has characterized the relationship between political discourse and the citizen is the term "Out-of-Bounds" (OB) markers, which refers to an invisible but well-established boundary in Singapore politics.

When a discourse is labeled unacceptable by the authorities, one is deemed to have breached the OB markers and stepped into a forbidden or unacceptable space and thus incurred the wrath of the authorities. Political discourse is held in check through regulations including authoritarian laws such as the Internal Security Act (ISA) which allows imprisonment without trial, as well as social norms such as the OB markers which sanctify politics as a sacred domain of the elite. The PAP's mix of ideologies leverages on the problematic concept of "Asian values" which constant-ly attaches essentialist qualities to Asian culture as a whole, to seek compliance and submission to its brand of democracy and politics. As such the OB markers are a referential cultural construct to imagine and limit the community's and the individ-ual's engagement with politics. The absent presence of this cultural taboo both in online and offline narratives plays a pivotal role in creating an intertextuality between the silent discourses in offline society and the alternative discourses on the Internet. In terms of the blogosphere, the individual agency that may be present in blogs stands in direct contrast to the political culture of Singapore.

The constant vacillation of the PAP's discourse between liberalizing the island and containing unacceptable political discussions has created an atmosphere of confusion for citizens where the lines between acceptable and unacceptable political discourse are sometimes blurred but never completely erased. Here the intertextuality between acceptable and banished discourses sustains the resonance of both themes in the collective psyche. In recent times there has been some increased space for nonestablishment (but not opposition) figures to express criticism of government policy through the forum pages of local newspapers although these concern details of policy rather than any fundamental challenges to the PAP's agenda or philosophy (Rodan, 2000: 175). More importantly, the relationship between "self" and "politics" is highly regulated and circumscribed by numerous laws and social norms, and it is within this context that blogs as a cultural tool mediate the political landscape of Singapore.

Media as a Feudalized Space

The government's ownership of the domestic mass media (both print and broadcasting) and its sustained silencing and taming of foreign media through enactments that target their advertising revenue in the region or hold them to account for their critical views through costly and highly-publicized defamation suits signify a political ritual to sustain the monologism of the PAP and to confine political discourse to spaces sanctioned by it. By depicting the foreign media as the "other" it frames these alternative viewpoints as an alien violation or intrusion.

This discourse of "Otherness" becomes another dimension in the politics of resistance on the Internet where "Otherness" is difficult to define and delineate in the online medium. The approach toward foreign media organizations has been "if you want to set a political agenda, then you have to be in the political arena. Otherwise you don't have the accountability and responsibility of looking after the place" (cf. Rodan, 2000). The notions of space and sovereignty become important criteria in narrating the nation, and as such "foreign voices" are depicted as exterritorial and illegitimate elements in narrating the nation-space. The discrediting of opposition politics and politicians in the domestic government-owned press and the constraints and strict vigilance on public assembly and speech over the years have further augmented the monologism of the PAP. The PAP's control of the spaces for political articulation creates a linearity in the political and historical narrative of the nation state. This visible monologism becomes a referential system in tracing the intertextuality between discourses that happen offline and in the unmediated spaces online and, equally, the resistance that can occur in such spaces.

There is a further qualification that needs to be made about the PAP-dominated public sphere. In the Habermasean public sphere (Habermas, 1964/1974: 49–50) "citizens behave as a public body when they confer in an unrestricted fashion—that is with guarantee of freedom of assembly and association and freedom to express and publish their opinions." In Singapore, although citizens may be critical of the government in their private lives, as signified through the common Singaporean phrase "coffee-shop talk," the sphere of social and political commentary, which occurs in the local media spaces is a highly policed sphere delineating both a linearity and historicity that thwart plurality as a guiding force in its representation. The evanescence of "coffee-shop talk" and the visible historical memory created by government-controlled media artifacts preserve the narratives of the nation from the perspective of the powerful (i.e., the PAP), creating this official sphere of political and social commentary which is not synonymous to a normative Habermasean public sphere.

The PAP's feudalized ownership of media spaces and its incessant ritualized silencing and compliance-seeking are part of its strategy to sustain this monologism despite the modernization and proliferation of media spaces. The "imagined nation" is, as such, constructed from the mythologizing of order and disorder in the spaces of governance. The media as a site of myth and memory sustain both an imagination of order and progress as well as the possible dangers which can derail this mythic order. Here "myth acts as a charter for the present-day social order; it supplies a retrospective pattern of moral values, sociological order; and magical belief, the function of which is to strengthen tradition and endow it with a greater value and prestige by tracing it back to a higher, better, more supernatural reality of initial events" (Brennan, 1990: 45). The PAP's rhetoric of transforming a "fishing village" into a twenty-first century metropolis leverages on the party's power to mythically transform a nation-space and equally on the dangers of losing these magical qualities if rules and moral values are not observed. The possibility that Singapore's achievements might be quickly and easily reversed is presented as a present danger that justifies the discriminate silencing of the nation-space as is the demand that strict protocols pertaining to political discourse be observed.

The Internet's Role in the State of Singapore

The Internet in its inception was received with a wave of euphoria, mainly due to its potential to reconfigure the nation state into an "intelligent island" wired through a new interactive technology (Mahizhnan & Yap, 2000). The economic imperative prompted the authorities to adopt a "light touch" in regulating the Internet. In 1996, the Internet code of practice was introduced and prohibited content which "is objectionable on the grounds of public interest, public morality, public order, public security, national harmony or is otherwise applicable by Singapore laws" (Media Development Authority—MDA Web site). The vagueness of what is acceptable on the online environment and the automatic imposition of the code on all the users, according to Terence Lee (2001a), was Foucauldian in nature for it sought to impose the subtle and existing forms of social control and conditioning evident in offline society on the virtual environment. According to Sussman (2003: 46) this "wide range of laws proscribing speech is intentionally vague so as to have the most chilling effect on political discussions held outside the gates of PAP-controlled channels." The authorities' employment of a combination of all four forms of control outlined by Lawrence Lessig (1999), that is, mediating the online environment via regulations, social norms, manipulation of the market, and the architecture of the Internet, meant that the approach to a large extent relied on both loosely defined new laws pertaining to the online environment as well as existing social norms regulating discourse in the republic.

The "code" sought to demarcate the "political" in the online environment and required political content providers to register with the Internet regulator, the MDA. This delineation of the political in the online environment has taken various forms over the years and has the intent of inhibiting the mobilization potential of political actors (Rodan, 2003). For example, when a political discussion group, Socratic Circle, posted survey questions to solicit opinions from Internet users in 1995, officials from the registrar of societies asked the group to discontinue reaching out to nonmembers through the Internet because they were contravening the rules of the Societies Act (Gomez, 2002: 3).

Prior to the 2001 elections the government announced further rules for using the Internet for political communication, and often these entailed curtailing or manipulating the use of features within the online environment. The new antielectioneering laws in 2001 outlawed messaging services over mobile phones on election content (Lee, 2002) although the Parliamentary Elections Act (PEA) in April in 2006 forbade the explicit streaming of political content by political parties and individuals. This meant political parties could not use podcasts or videocasts for election advertising. Additionally, the authorities also warned that "bloggers who persistently propagate, promote or circulate political issues relating to Singapore will be required to register with the Media Development Authority (MDA) and once registered these bloggers will not be able to provide material online that could constitute election advertising" (www.mica.gov.sg). Those who contravene the law can face a jail sentence of up to 12 months, a S$1,000 fine, or both.

The authorities also prohibited nonparty Web sites from carrying party-related material (Tsang, 2001). This is consistent with offline policies, which discourage alliances between political parties and civil society organizations, often extracting a form of nonpartisan politics from nonparty political entities in Singapore. This has in many ways resulted in civil society organizations professing themselves to be nonaligned entities to ensure their existence or survival in the political landscape. This has invariably strangled and thwarted the support base of political parties. As such, the online restrictions were targeted at political entities which may be a potential threat to existing power relations in Singapore. The regulatory environment sought to extend the offline rules of political discourse onto the Internet.

The Blogosphere and the General Election

In Singapore, over 67 percent of the island's 4.4 million population is wired to the Internet, and its Internet penetration has been ranked by Internetworldstats.com as the third highest in the world after Japan and Hong Kong (Crispin, 2006). This has meant that seven in ten homes have Internet access and seven in ten Internet users are bloggers (Feng, 2007) According to a survey by the Infocomm Development Authority of Singapore released in March 2007 there are an estimated 862,000 bloggers and blog readers in Singapore, and this survey estimated the blogging population to be 502,000 in 2006. In addition, at least 21 politicians both from the PAP and opposition parties have set up blogs (Feng, 2007).

In Singapore, blogs have come under scrutiny in different ways. In September 2005, three bloggers were charged under the Sedition Act (which dates back to the colonial era) for posting anti-Muslim remarks on their blogs. These remarks were deemed as threatening the social harmony and political stability of the island state. International press freedom group Reporters Without Borders described the lawsuit as "intimidation" that "could make the country's blogs as timid and obedient as the traditional media" (*Taipei Times*, September 19, 2005). In November 2005, the government barred military servicemen from posting images and text on army life. It justified its decision on the grounds of security (*Sydney Morning Herald*, November 21, 2005). In another incident a University of Illinois student from Singapore was threatened with a lawsuit over comments made on his Web journal about a government-linked research agency. The research unit agreed not to sue after the student shut down his blog and apologized for the comments he posted about the agency (*South China Morning Post*, August 18, 2005). These separate incidents highlight some of the social issues new cultural practices such as blogging can create in a society such as Singapore.

There is often a link between mediatized events and the proliferation of citizen journalism, as in the case of the Asian tsunami, 9/11, and the war in Iraq. Similarly in Singapore, "from the time the writ of election was issued on April 20, 2006, there was a flurry of posts appearing on blogs and forums" (Lee, 2006). The Election Department had also warned that the blogosphere was being tracked (cf. Siew & Chua, 2006). A report by the Institute of Policy Studies revealed that up to 50 Web sites and blogs had political or "semi-political" material during elections. But despite this, political blogging during the election did not stop. According to Technorati.com, an independent blogger search engine, the names of three Singaporean bloggers ranked among the world's top ten used search words (Crispin, 2006).

Reports in several government-owned newspapers confirmed that there was a rise in the number of blogs dedicated to election commentary as well as an increase in traffic to these Web sites (see Lee, 2006; Siew & Chua, 2006). Prior to the announcement of the amendment to the Parliamentary Elections Act, for example, "new blog entries with the words 'Singapore Election' ranged between 12 and 30 per day with more than 100 new politically-oriented entries uploaded on some days in March" (Crispin, 2006). Various terminologies have been applied to the alternative political content occurring on these virtual platforms. Cherian George (2003: 1) terms it as "politically contentious journalism" which challenges the dominant ideologies as it "attempts to democratize political discourse." Additionally, the blogosphere is dispersed and makes enforcement of regulations less effective (Au, 2007).

One important aspect of the blogosphere is that it showcases the convergence of various streaming technologies, thus the discursive sphere is often complemented by sounds and visuals along with commentary from individuals and other sources such as newspaper reports. The links also join together other Web spaces, a process which widens the political commentary beyond any given platform. In Singapore, in addition to the proliferation of Web sites and blogs dedicated to election commentary, dozens of rally videos emerged on the Internet, and these were mainly images captured by citizens through hand-held cameras and mobile phones and were uploaded through technologies such as YouTube and Google video (cf. Siew & Chua). Although political parties were banned from using podcasts and videocasts under the amended Parliament Act, these blogs publicized and made the podcasts and videocasts of rallies available online. When the rallies began on April 28, 2006, dozens of short video clips capturing the rallies appeared on Web sites such as SG Rally and Singapore Election Rally videos. One local blog (www.singapore-govt.blogspot.com) received between 5,000 and 6,000 hits during the nine-day election campaign period, double what it usually gets (Lee, 2006).

As Tan Tarn How of the IPS points out "most blogs are anonymous and difficult to track and even if these were tracked and clamped down it could lead to a negative backlash for the authorities during the campaign period" (cf. Lee, 2006).

In addition, blogosphere researchers such as Tan Tarn How concede that these Web sites provided "good analysis where the postings helped clarify and increase knowledge" (cf. Lee, 2006). The anonymity of the blogs and the participatory element where people were able to send in their accounts and visuals of rallies represented a discursive sphere which was a counterpoint to the government-managed public sphere of offline society.

A postelection survey commissioned by the IPS found that so-called "bread and butter issues" were not the main concern of voters, contradicting received wisdom that Singaporeans are a pragmatic lot. The desire for an efficient government and fairness of government policy, including the wish for a plurality of views on politics and other issues, was found to be a matter of importance (Asian Analysis). The IPS postelection survey also indicated that the Internet as a source of election-related information still lost out to other mainstream media, as well as informal information sources such as word of mouth and door-to-door visits by candidates (Kwek, 2006a). The political commentary of the blogs nevertheless renegotiated the "survival discourse" that is often employed by the PAP, where economic survival and a compliant political culture are seen as imperative to the sustained growth of the nation. The extension of discourse beyond "bread and butter" issues is indicative of the plurality of perspectives that can emerge on the blogosphere.

Nexlabs, a local Web site tracing firm which had been monitoring the blogging scene since January 2006, counted over 2,000 election-related postings and the most popular discourses included debates about an open society, defamation suits, education, and the National Kidney Foundation scandal (Siew & Chua, 2006). According to NexLabs, despite rules to caution bloggers about persistent political content "people have not stopped blogging and they are starting up new topics all the time" (cf. Siew & Chua). In the recent years there has also been a proliferation of more public and social ones involving more blog aggregators and group blogs where a collection of personal blogs is collated into a single site in the former and a team of authors are involved in the latter (Feng, 2007). The inclusion of the masses creates a broader spectrum of viewpoints as well as readership. Aggregating sites collect entries from contributors' personal blogs, generating not only a greater readership but often forming communities and linkages with other blogs.

Popular blogs such as Ping.sg. Intelligent Singaporean, and Singapore Angle receive about 100,000, 45,000, 50,000 page views per month, respectively. Some blogs such as Mr. Brown and Xiaxue have gained mainstream acceptability and recognition by being mentioned and reported about in the local newspapers. An example is the Derek Wee incident where a blogger penned his insecurities about growing old in Singapore. When Wee Shu Min, the 18-year-old daughter of an MP, criticized Mr. Wee as lazy in her blog, the story was widely circulated in the blogosphere. "Her attack was criticised by hundreds of Internet users who accused her of being elitist, naïve and insensitive to the lives of Singaporeans from humbler backgrounds" (Kwek, 2006b).

The incident became an iconic moment for the bloggers when the MP Wee Siew Kim apologized to the nation for his and his daughter's comments in the mainstream media (see Kwek, 2006b). The incident received national coverage in the press and equally the blogosphere was filled with comments about the incident, and many bloggers interpreted it as politics submitting to popular opinion as expressed through blogs. Beyond commentaries multimedia productions satirizing Wee Shu Min's allegedly elitist comments also circulated on the Internet. The relationship of blogs to the mainstream media is an important element that builds an intertextuality that sustains everyday discourses of resistance and is discussed further in the chapter in relation to the concept of "intertextuality."

During the recent elections, in contrast to the PAP's ideology of politics as a serious issue, humor mediated political discourse in blogs. Popular bloggers Lee Kin Mun and Benjamin Lee, who are, respectively, known in the virtual sphere as Mr. Brown and Mr. Miyagi, produced a podcast in which they spoofed Worker's Party Candidate James Gomez, who had a run-in with the government over his application to be a minority candidate. The spoof was downloaded more than 30,000 times in the three days after it went live and "became the most talked about piece of political satire during the recent election" (Koh, 2006). According to Minister for Information, Communications and the Arts, Lee Boon Yang, the audio clip was "clever and funny but may not contribute to a better understanding of deep-seated issues" (Koh, 2006).

The blogosphere has also thrown up personalities who have had an impact on offline politics in Singapore. When Lee Kin Mun's weekly column in the free paper *Today* was suspended for being critical of the government, 30 people gathered in the City Hall train station to protest against the suspension (*Straits Times*, July 22, 2006). In the article entitled "Singaporeans are fed up with progress," Lee Kin Mun had commented on the high cost of living in Singapore. The prime minister commented on the article in his National Day Speech on August 21, 2006, reiterating that "this is not the way to carry on a public debate on national issues, especially not in the mainstream media," thus highlighting both the political economy of the mainstream media as well as the lack of a robust public sphere in Singapore.

The protestors responded to an SMS request to protest against the decision by wearing brown and convening at the specified location. In the column, Mr. Brown had commented on the hikes in taxi fares and electricity bills which had been implemented after the polls and at a point when a government survey had revealed a widening income gap between rich and poor (*Straits Times*, July 22, 2006). This demonstrated that blogs are creating new forms of political personalities beyond those of politicians elected through the electoral system or the leadership of the PAP.

The blogs also played a role in widening the viral appeal of a petition which called for the resignation of the CEO of a government-backed charity, the National Kidney Foundation (NKF), when the organization was deemed to have misappropriated public donations. The public outcry against the incident was unprecedented with online forums and the blogosphere inundated with opinions and criticism. An online petition started by an Internet user received 40,000 signatures, and several blogs played a role in both drawing the attention of their readers to the issue and encouraging them to show their discontent through the petition. The public outcry led to the resignation of the NKF CEO and its board and patrons, and the government assured the public that the matter would be fully investigated (see Seah, 2005).

Despite government criticism of the Internet as an unreliable space with hidden agendas, the PAP recognizes that with growing media literacy and access due to its own information and technology policies, the Internet will be increasingly used by the younger generation to engage in political discourse. In a poll conducted by the MDA in 2006, half of all teens between the ages of 15 and 19 maintained a Web log compared to 46 percent in the 20–24 category (Au, 2007). Consequently PAP MPs are keen to engage younger voters from the post-65 generation (i.e., post-independence) who made up nearly 40 percent of the electorate in the 2006 general election. It has declared recently that it "may need to reach out to these voters through new media, podcasts and blogs" (*Channel New Asia*, June 5, 2006). This is also partly in recognition of the fact that many of the blogs were anti-government during the election period. According to MP Denise Phua "something has gone wrong when more than 85 percent write negatively about the PAP. The government should figure out how to manage this channel of communication" (cf. Au, 2007).

Anonymous Websites and Territorial Politics

In contrast to offline society, the notion of "otherness" is a much more tenuous concept in the online environment and may not be strategically used as a boundary marker as it is in the offline environment. The ability of technology to create new personas as well as strip and reinvent identities means that identity on the Internet cannot be neatly divided into "insiders" and "outsiders" or "citizens" and "foreigners." In the online environment, political commentary from anonymous Web sites falls into a sphere of ambiguity due to this "avatarism" (Judith, 1998). The growth of anonymous Web entities based in unverifiable locations and dabbling in political discourse pertaining to Singapore is a well-documented phenomenon (George, 2006; Ho, Barber, & Khondker, 2002; Lee & Wilnat, 2006; Sussman, 2003; Tan,

2006:15). Unlike the local political Web sites which are mediated by regulations in Singapore, these Web sites operate under the cloak of anonymity and are playing an important role in expanding the sphere of political discourse in the online environment (Gomez, 2006; Lee & Wilnat, 2006).

In the last two elections (2001 and 2006), anonymous Web sites have played a role in widening the debates and in carrying political material on opposition political parties in Singapore. The appropriation of election commentary by anonymous Web sites challenges the trajectory of domestic media and party apparatuses having a monopoly on election commentary in Singapore. Although nonparty Web sites are forbidden from carrying party material, these anonymous Web spaces publicize party material and discourse and often highlight the plight of the opposition in Singapore (Gomez, 2006). In the narratives of the anonymous Web sites, the victims of Singapore politics become the new protagonists. The tendency to have anti-establishment discourses in the online medium, although consistent with the findings of Hill and Hughes (1998), does nevertheless inverse the rules of political discourse in Singapore, thus expanding spaces for opposition politics and presenting issues from their perspectives. Anonymous Web sites as such have de-centered political narratives in the virtual sphere.

The politics of control and the regulatory environment lead to the formation of a two-tier system of control in the online environment. The first sphere is one that consists of political Web sites (including those belonging to political parties and civil society organizations) which are located in Singapore and amenable to control via the regulations imposed on them. This sphere of commentary is one which can be directly mediated by the government, and hence it is amenable to rules of political discourse and regulations imposed by the context of its embedding.

In comparison the second tier, comprising anonymous Web sites and blogs, is distinct from the first in that it is less amenable to direct regulatory control. Consequentially, this sphere is able to appropriate discourses which are banished or marginalized in offline society. It has expanded the sphere of commentary to accommodate topics and discussions deemed too sensitive or taboo by the authorities. Unlike foreign media firms, which are economic entities that can be held accountable in a physical space, these anonymous Web sites can mushroom, relatively unregulated, in the virtual sphere. These anonymous Web sites, by engaging with the politics of Singapore, have renegotiated the established rules of political discourse. Such intertextuality needs to be viewed as a structural analysis of texts in relation to the larger system of signifying practices or users of signs in culture (Landow, 1992: 10–11).

The expansion of spaces for opposition politics and alternative discourses makes this second sphere a crucial element in the contemporary politics of Singapore, both materially and symbolically. Although it offers a form of symbolic, textual, and visual resistance to social norms and discourses discouraged or displaced from offline society, it offers a sustained counterdiscourse to the politics in the island, which hitherto had been limited due to the government's degree of control over the domestic and foreign media. The posting of information between the controlled sphere and the anonymous one (Gomez, 2002, 2006; Lee & Wilnat, 2006) creates new forms of virtual community and fraternity. The notion of intertextuality liberates the text from psychological, sociological, and historical determinisms, opening it up to an apparently infinite play of relationships (Landow, 1992).

These two tiers of political commentary are not hermetically sealed. Both spheres of commentary are interlinked through discourses which happen in offline society. The intertextuality between the two tiers and offline society happens through the topics of discussion, forming a referential system to sustain the connections. Another linguistic framework is the use of Singlish (i.e., Singaporean English) and local humor which form points of identification between the offline community and the virtual space.

Intertextuality and Hypertext

As mentioned earlier, a distinct feature of blogs is that they are link-heavy, which emphasizes the need to consider the decentering of information and forms of connectivity or networks they can create. The hypertext system enables the reader to branch off from what appears to be a main text into intertextual pathways, to the extent that the main text may be forgotten or come to seem just one more text in an intertextual or hypertextual chain (Allen, 2000: 202). One of the distinct characteristics of hypertext is that it is composed of bodies of linked texts that have no primary axis of organization. This ability to decenter and recenter is a distinct characteristic of the online digital environment. Landow (1992), in quoting Derrida, points out that "de-centring has played an essential role in intellectual change for a dislocation from the culture of reference is a vital function of this decentring." On the Internet, text becomes unbound and nonlinear in its reading. Hypertext, as Landow (1994) intimates, makes author, text, and reader into joint participants of a plural, intertextual network of significations and potential significations (Allen, 2000: 202).

With reference to the blogosphere in Singaporean politics, besides aggregated blogs the connections between offline issues and online discourses are created through hypertext and URL links within blogs which often signpost similar blogs and wider issues and debates happening in other blogs. This is evident in blogs such as Singabloodypore, The Online Citizen, Singapore Surf, and the Intelligent

Singaporean among others. The streaming of information in the blogosphere that is banned in political party Web sites also creates new forms of virtual communities and fraternities (Gomez, 2002, 2006; Lee & Wilnat, 2006).

The intertextuality between the online sphere and offline society happens through the topics of discussion, forming a referential system to sustain the connections linking the government-managed public sphere with the plural and banished discourses of the offline platform. Often intextuality with offline society is created through articles or reports published in the mainstream media. Various discussions in blogs often premise on mainstream media articles as points of enquiry highlighting both what has been conveyed or what has been omitted by the mainstream media. In a recent incident, popular blogger Alex Au raised doubts about the statistical figures concerning the price of government-subsidized flats in a *Straits Times* article in July 2007 in his Web site YawningBread.org. The doubts raised by Alex Au were covered by another local paper, *Today*, and subsequently the relevant body was forced to issue a statement that it had published inaccurate information from an outdated source (see Kwok, 2007). Although the mainstream newspapers are used as a platform for critique, they can also provide a degree of validation by the bloggers. The *Straits Times*' coverage of a bloggers' conference in July 2005 created many reviews in the blogosphere, and various bloggers viewed it as being recognized by the mainstream media and offline society.

The intertextuality between the offline media and society on the one hand and the blogosphere on the other can be also discerned through the implicit counterdiscourses and counterpoints which are raised in the blogosphere. The counterpoints in this sense may be evident to those familiar with the resonant discourses in Singapore society. This is evident in the juxtaposition of ideas or themes that can occur in dissident Web sites. For example when the prime minister mooted the idea of people sending in pictures of them smiling so that the nation-state could welcome the delegates convening for the International Monetary Fund (IMF) and World Bank (WB) meeting in Singapore in September 2006 with 4 million smiles, the blogsosphere responded to the theme by creating various types of discursive and multimedia material as a counter to this suggestion. The Mr. Brown blog, for example, contained a podcast which satirized the delegates coming into a land where people only smiled, and in another podcast in the same Web site there is a parody on Singaporeans being taught how to be good protestors which was obviously aimed at the government's clampdown on street protests during the IMF/WB meeting. Similarly, a 19-year-old student set up http://4millionfrowns.livejournal. com asking people to send in their frowning photographs as a counterpoint to the government's promotional campaign and to express their discontent with the government barring entry to social activists and nongovernmental organizations (see Ibrahim, 2006).

The intertextuality between the mainstream media and blogs is bound with the visible (i.e., what has been presented by the mainstream media), the absent (i.e., what has been omitted) and counter discourses which raise sustained arguments against the prevailing ideologies without necessarily referring to them. During the 2006 elections, various commentaries in the blogosphere were focused on bringing to the public what had been omitted by the mainstream media. One prominent example is the widely circulated image dubbed as the "Hougang photo" on the Yawning Bread Web site which encapsulates an image of people thronging to hear the election rally of an opposition party. The photo captures the image of thousands of people attending the Worker's Party rally at Hougang town, and the resounding support of the turnout stands in contrast to the media portrayal of the population as politically apathetic and opposition parties as noncredible entities.

The political economy of the mainstream media and the dominant political ideology provide a canvas for creating a counter discourse in the blogosphere. This means that what is controversial or novel for the community can be influenced by offline societal norms. In the blogosphere, in particular, what can qualify as a "media event" can be intimately influenced by issues of visibility or absence in the mainstream media. These can include extra coverage and first-hand accounts of street protest or demonstrations. The Online Citizen, for example, reported on a protest march in September 2007 by the Singapore Democratic Party to commemorate a protests which was prevented by the authorities a year ago. The Web site promised its readers to post live updates and pictures from the march. These media events created from the vantage point of the citizen renegotiate issues of the temporal and the concept of "live" events in comparison to the mainstream media. Equally the use of humor, sarcasm, parody, and satire to discuss politics challenges the offline rhetoric of framing politics as a serious matter.

The complex Web of intertextuality between the mainstream media and the blogosphere does not completely renegotiate the issue of center and periphery when discussing traditional and new media, but it does provide new platforms to reframe dominant conceptions of what could constitute "political." In the process it creates a mediated memory in the blogosphere where archived postings and commentaries create a trajectory of what is political, often drawing parallels and analogies through what has been narrated in these spaces. A case in point is the Otto Fong incident in which Mr. Fong, a teacher, had posted a letter about his sexuality in his personal blog. The incident had received media attention in the local press but soon afterward it was taken down. This prompted much commentary in the blogosphere where his original entry has been archived for the audience interested in reading it. Second, many of the blogs referred to an earlier incident in which a prominent playwright had been fired from a relief teaching position by the Ministry of Education. There was speculation in the blogosphere that this could have been prompted by the political leanings of his plays which have homosexual themes and an antigovernment stance.

Reading Forms of Empowerment in the Blogosphere

Au (2007) points out that, unlike the situation in Malaysia and South Korea, the audience for political Web sites and blogs in Singapore remains small and atomized, and thus despite the vibrancy of the blogosphere there is no critical following of it. The post-election IPS survey also confirmed that television and print media were the main vehicles of information. Nevertheless, there is a need to consider the role of new information and communication technologies (ICTs) beyond comparing them with embedded traditional media. It may be more pertinent to comprehend the interconnectivities within the ecology of media, where new forms of resistance and counter discourse can flourish without necessarily offering a degree of "centredness" or critical mass that mainstream media may provide.

The blogosphere, as a platform which showcases a plethora of individual and communal voices, thwarts the linearity of discourses which the government has carefully managed and cultivated over a number of years. Unlike speeches, which are evanescent, the Internet as a textual and visual medium embodies a different type of utterance—one that is visually present but de-centered due to the sheer volume of material available on the Web. In addition, the Internet suffers from a crisis of credibility as it is a receptacle for both false and valid information (Katz, 1998). Although the foreign media, whether print or online, have been taken to task by the government, the alternative views found on the Internet add a different dimension to the rules of political discourse in Singapore. They incorporate a dialogism between the enforced and unfettered—conjoining valid commentaries with the salacious and noncredible—and in the process form a Web of articulations that resist and remove the sanctity and the sanitization associated with political discourse in Singapore.

The convergence of different genres and technologies in the blogosphere has "de-feudalized" the government-dominated public sphere, expanding the terrain to accommodate critical voices, dissent, and individual authorship. In the process they are renegotiating the relationship between self, civic participation and politics but more specifically the tacit rule of Out-of-Bounds markers. The visual and textual terrain of the Internet also celebrates and consolidates performative elements as "the very presence of utterance is historically and socially significant" (Bakhtin & Medvedev, 1978: 120; cf. Allen, 2000: 17). No word or utterance from this perspective is ever neutral, and, as such, the Internet highlights the linguistic interactions of specific individuals or groups within special social contexts. Inevitably, they also record civil and social history differently from the dominant discourse of the ruling party.

The conjoining of forbidden and acceptable discourses online provides a conundrum for the authorities, making the Internet a double-headed hydra. The Internet is both a regulatory challenge and a medium which cannot be completely ignored. The general election in 2006 inspired a Bakhtinian "carnivalesque" atmosphere on the Internet, with blogs blatantly transgressing the ban on audio and video streaming of election rallies. The parade of political satires, caricatures, and citizen reports represented a carnival of resistance to established rules and norms. As a form of ironic representation, parody is doubly coded for it both legitimizes and subverts that which is produced (Hutcheon, 1989: 97).

In Bakhtin's (1984) "carnival act," the unofficial dimensions of society are manifest through profane language and drama. Through these depictions the carnival celebrates the unofficial collective body of people who stand against official ideology and state power and in the process overturn the dominant order of the society. Although Bakhtin's writing is in reference to the novel as the artifact of subversion, the blogosphere in the context of Singapore offers a platform of resistance where rules are renegotiated or even subverted. The language of monologism set out by the PAP for the domestic mass media is challenged by the Internet, a performative arena of anarchy and rule-breaking (Allen, 2000: 30).

The ability to remove one's identity and yet retain one's voice and sense of presence in the blogosphere signifies both the presence and absence of the self in the online environment. This disembodied presence again presents a new dimension in Singaporean politics where the "self" mediated through technology can form a different relationship with politics. Blogs are a new media genre that inhabit a problematic space between the personal and the public, and invariably anonymous blogs celebrate both the elements of anonymity as well as the creation of a signature voice in the blog space. The personal and anonymous may be dialectical but they nevertheless emphasize the ability of individuals to participate in political commentary. Singapore-based blogs have also created a new culture where nonpolitical personalities have a presence in the political domain through their blog commentaries. The use of humor in many blogs also re-negotiates the rules of political discourse, creating different ways of delivering and engaging with politics where there is a subversion of established norms.

Media reports have identified the growing presence of an Internet-savvy younger generation of Singaporeans contributing to political blogging to express their views (Chia, 2006) and undeniably the government can control the extent of the discourse through its architecture of old and new regulations. Nevertheless, the Internet as a de-centered space accommodates various kinds of discourses and activities enabling new links and fraternities to be established in cyberspace. The articulation of opposition party politics in nonparty political Web sites transgresses the rule of nonpartisan politics often enforced by the PAP. The intertextual links between party and nonparty anonymous Web sites on the Internet create a new communal online politics which highlights new forms of communion between virtual and real entities (i.e., political parties, NGOs, and civil society organizations) with an interest in Singapore politics. In this context the Internet is enabling new political groupings (Cleaver, 1998) to create a type of public sphere very different to that found offline. These spaces are also contributing to a certain degree of political socialization and perhaps re-socialization in Singapore, where people learn to participate in a technologically mediated politics.

Conclusion

The politics of resistance in the blogosphere constitutes a sphere where norms and rules are continually challenged, inversed, or even transgressed. The virtual is not a dichotomous sphere divorced from the physical, and intertextuality binds these spaces in obvious and complex ways. The Internet, as an electronic space, which enables new forms of expression, identity, and embodiment, has to nevertheless reconcile the social and political context of its physical embedding (Miller & Slater, 2000). The types and degrees of empowerment enabled by virtual politics can only be traced, discerned, and understood through this intertextuality between the physical and virtual. These intrinsic and incestuous connections between the online and offline environments provide the basis to interpret the degree and types of empowerment (i.e., whether symbolic or material) that can occur with textual and visual online resistance.

Bibliography

(2006, April 3). *Podcasting not allowed during elections*. Retrieved June 6, 2006, from Channel News Asia Web site. http://www.channelnewsasia.com/stories/singaporelocalnews/view/201330/1.html

(2006, June 5). *Engaging younger voters key to meeting future aspirations: New PAP MPs*. Retrieved June 6, 2006, from Channel News Asia Web site: http://www.channelnewsasia.com/stories/singaporelocalnews/print/212077/1/.html

(2006, July 22). No action against Mr Brown's supporters. *The Straits Times*.

Abdul, R. F. *Blogging activity up during Singapore election campaigning/video*. Retrieved May 12, 2006, from Channel News Asia Web site: http://www.channelnewsasia.com/stories/singaporelocalnews/view/207967/1/.html

Allen, G. (2000). *Intertextuality*. London: Routledge.

Anderson, B. (1991). *Imagined communities: Reflections on the origin and spread of nationalism.* London: Verso.

Ang, P. H., & Nadarajan, B. (1996). Censorship and the internet in Singapore. *Communications of the ACM, 22*(1), 99–114.

Asian Analysis (2006, August). *'Podcasting Not Allowed During Elections,'* Channel News Asia, April 03 2006. Retrieved June 6, 2006, from http://www.channelnewsasia.com/stories/singaporelocalnews/view/201330/1.htmlPost-election blues. Retrieved April 12, 2007, from http://www.aseanfocus.com/asiananalysis/article.cfm?articleID=974.

Au, A. (2007). *To blog or not to blog in Singapore.* Retrieved June 22, 2007, from Asia times Online Web site: http://www.atimes.com/atimes/Southeast_Asia/IC10Ae01.html

Baber, Z. (2002). Engendering or endangering democracy? The internet, civil society and the public sphere. *Asian Journal of Social Science, 30*(2), 287–303.

Baber, Z. (2003). New media, new politics? The internet and the prospects for digital democracy. *Bulletin of Science, Technology and Society, 23*(2), 125–128.

Bakhtin, M. M. (1981). *The dialogic imagination: Four essays.* In M. Holquist (Ed.), Emerson, C., & M. Holquist (Trans.). Austin, TX: University of Texas Press.

Bakhtin, M. M. (1984). *Rabelais and his world.* In C. Emerson & M. Holquist (Eds.), I. Helen (Trans.). Austin, TX: University of Texas Press.

Bakhtin, M. M., & Medvedev, P. N. (1978). *The formal method in literary scholarish: A critical introduction to sociological poetics* (J. W. Albert, Trans.). Baltimore, MD and London: Johns Hopkins University Press.

Banerjee, I., & Yeo, B. (2003). Internet and democracy in Singapore: A critical appraisal. In I. Banerjee (Ed.), *Rhetoric and reality: The internet challenge for democracy in Asia.* Singapore: Eastern Universities Press.

Barthes, R. (1976). *Pleasure of the text* (Richard, M., Trans.). New York: Hill and Wang.

Bhabha, H. K. (1990). DissemiNation: Time, narrative, and the margins of the modern nation. In H. K. Bhabha (Ed.), *Nation and narration* (pp. 291–322). London: Routledge.

Blanchard, A. (2004). Blogs as virtual communities: Identifying a sense of community in the Julie/Julia project. In *Into the blogosphere: Rhetoric, community and culture.* Retrieved June 6, 2006, from http://blog.lib.umn.edu/blogosphere

Bokhorst-Heng, W. (2002). Newspapers in Singapore: A mass ceremony in the imagining of the nation. *Media, Culture & Society, 24*(4), 559–569.

Brennan, T. (1990). The national longing for form. In H. K. Bhabha (Ed.), *Nation and narration* (pp. 44–70). London: Routledge.

Burnett, R., & Marshall, D. P. (2003). *Web theory.* London: Routledge.

Chan, H. C. (1976). *The dynamics of one-party dominance: The PAP at the grass-roots.* Singapore: Singapore University Press.

Chia, S. (2006). New media, same rules. *The Straits Times.* Retrieved June 6, 2006, from http://www.asiamedia.ucla.edu/print.asp?parentid=43361

Chin, A., & Chignell, M. (2006). *A social hypertext model for finding community in blogs.* In Proceedings of the 17th ACM Conference on Hypertext and Hypermedia: Tools for Supporting Social Structures, Odense, Denmark, August 23–25, 2006.

Christie, K. (1997). Fear is the key: Singapore's brave new world. *Australian Journal of Political Science, 32*(1), 123–129.

Chua, B. H. (1995). *Communitarian ideology and democracy in Singapore.* London: Routledge.

Chua, B. H. (1996). Culturalisation of economy and politics in Singapore. In R. Robison (Ed.), *Pathways to Asia: The politics of engagement.* St. Leonards, NSW: Allen and Unwin, 87–107.

Cleaver, H. M., Jr. (1998). The Zapatista effect: The internet and the rise of an alternative political fabric. *Journal of International Affairs, 51*(2), 621–640.

Crispin, S. W. (2006). *Singapore's authoritarian rulers tangled in web.* Retrieved June 14, 2007, from Asia Times Online Web site: http://www.atimes.com/Souteast_Asia/HD27Ae03.html

Deleuze, G., & Guattari, F. (1987). *A thousand plateaus: Capitalism and schizophrenia* (B. Massumi, Trans.). Minneapolis: University of Minnesota Press.

Deleuze, G., & Guattari, F. (1993). Rhizome versus trees. In C. V. Boundas (Ed.), *The Deleuze reader* (pp. 27–36). New York: Columbia University Press.

Derrida, J. (1995). *Archive fever* (E. Prenowitz, Trans.). Chicago: University of Chicago Press.

Eco, U. (1986). *Travels in hypereality* (W. Weaver, Trans.). San Diego: Harcourt.

Eco, U. (2003). *Mouse or rat: Translation as negotiation.* London: Weidenfeld and Nicolson.

'Engaging Younger Voters Key to Meeting Future Aspirations: New PAP MPs,' Channel News Asia, June 5, 2006. Retrieved June 6, 2006 from http://www.channelnewsasia.com/stories/singaporelocalnews/print/212077/1/.html

Everett, A. (2003). Digitextuality and click theory: Theses on convergence media in the digital age. In *New media: Theories and practices of digitextuality* (pp. 2–28). London: Routledge.

Feng, Z. (2007, July 17). More citizens open up to blog scene. *The Straits Times.* Retrieved September 17, 2007, from http://digital.asiaone.com/Digital/Features/Story/A1Story20070720–19096.html

George, C. (2001). *Singapore: The air-conditioned nation.* Singapore: Landmark Books.

George, C. (2003). *The internet's political impact and penetration/participation paradox in Malaysia and Singapore* (Working Paper No. 14). Singapore: Asia Research Institute, NUS.

George, C. (2006). *Contentious journalism and the internet: Towards democratic discourse in Malaysia and Singapore.* Singapore and Seattle: Singapore University Press and Washington University Press.

Gomez, J. (2002). *Internet politics surveillance and intimidation in Singapore.* Singapore: Think Centre.

Gomez, J. (2006). *Democracy and internet in Singapore: The supply of alternative political content during general elections.* Paper presented at Imagining/Nation Without Borders in Bangkok, Thailand, 28–29 July 2006.

Habermas, J. (1964/1974). The public sphere: An encyclopaedia article. *New German Critique, 3*(1), 14–21.

Herring, S. C., Inna Kouper, John C. Paolillo, Lois Ann Scheidt, Michael Tyworth, Peter Welsch, Elijah Wright, and Ning Yuet al., (2005). *Conversations in the blogosphere: An analysis from the bottom.* Proceedings of the Thirty-Eighth Hawaii International Conference on System Science, Los Alamitos, pp. 107–118.

Hill, K. A., & Hughes, J. E. (1998). *Cyberpolitics: citizen activism in the age of the internet.* Lanham, MD: Rowman & Littlefield Publishers.

Hine, C. (2000). *Virtual ethnography.* London: Sage Publications.

Ho, K. C., Barber, Z., & Khondker, H. (2002). Sites of resistance: Alternative websites and state-society relations. *British Journal of Sociology, 53*(1), 127–188.

Ho, K. C., (Ed) Kluver, R. and C.C. Yang (authors). (2003). *Asia.com: Asia encounters the internet.* London: Routledge.

Hutcheon, L. (1989). *The politics of postmodern*. London: Routledge.

Ibrahim, Y. (2006, December). Weblogs as personal narratives: Displacing history and temporality. *M/C Journal, 9*(6). Retrieved June 23, 2007 from http://journal.mediaculture.org.au/0612/08-ibrahim.php

Judith, S. D. (1998). Identity and deception in the virtual community. In P. Kollack & M. A. Smith (Eds.), *Communities in cyberspace*. London: Routledge, 29–59.

Katz, J. E. (1998). Struggle in cyberspace: Fact and friction on the World Wide Web. *Annals of the American Academy of Political and Social Science, 560*, 194–199.

Khong, C.-O. (1995). Singapore: Political legitimacy through managing conformity. In M. Alagappa (Ed.), *Political legitimacy in South East Asia*. California: University Press, 108–35.

Koh, A., Ng Ee Soon, Benjamin H. Detenber, Mark Cenite. (2005). *Ethics in blogging* [Report Series]. Singapore: Singapore Internet Research Centre.

Koh, L. (2006, June 1). Net spoof too funny for serious politics? *The Straits Times*.

Kristeva, J. (1984). *Revolution in poetic language* (M. Waller Trans.). New York: Columbia University Press.

Kuo, E. C. (1983). *Communication policy and planning in Singapore*. Honolulu: Kegan Paul International in association with East West Institute.

Kwek, K. (2006a, June 3). Internet not yet a viable new media. *The Straits Times*. Retrieved June 6, 2006, from http://www.asiamedia.ucla.edu/print.asp?parentid=47115

Kwek, K. (2006b, November 24). Teen blogger counselled for her elitist remarks. *The Straits Times*. Retrieved September 17, 2007, from http://mrlim.isthebest.net/2006/10/24/wee-shu-min-violated/

Kwok, W. (2007). *Blogger reveals* Straits Times' *error in rental report*. Retrieved September 17, 2007, from AsiaMedia Web site: http://asiamedia.ucla.edu/article.asp?parentid=74640

Landow, G. P. (1992). *Hypertext: The convergence of contemporary critical theory and technology*. Baltimore and London: John Hopkins University Press.

Landow, G. P. (Ed.). (1994). *Hyper/text/theory*. Baltimore and London: Johns Hopkins University Press.

Lee, L. (2006, May 9). Net was abuzz with politics during poll period. *The Straits Times*.

Lee, T. (2001a). Auto-regulating new media. *Australian Journal of Communication, 28*(1), 43–56.

Lee, T. (2001b). The internet in Singapore: From self-regulation to auto-regulation. *Communications Law Bulletin, 19*(4). 1–5.

Lee, T. (2002). The politics of civil society in Singapore. *Asian Studies Review, 28*, 97–117.

Lee, T., & Wilnat, L. (2006). *Media research and political communication in Singapore* (ARC Working Papers No. 130). Retrieved October 1, 2006, from http://wwwarc.murdoch.edu.au/wp/wp130.pdf

Lemon, S. (2006, April 12). *Will Singapore's ban on political blogs work?* Retrieved June 6, 2007, from ComputerWorld Web site: http://www.computerworld.com/securitytopics/security/privacy/story/0,10801,110431,00.html

Lessig, L. (1999). *Code and other laws of the cyberspace*. New York: Basic Books.

Low, L. (2001). The Singapore developmental state in the new economy and polity. *The Pacific Review, 14*(3), 411–441.

Luo, S., & Yap, S. (2005, August 2). Online petitions: New wave in local social mobilisation? *Digital Life*. Retrieved June 22, 2007, from http://ww.ncs.com.sg/media/clippin.aspID=732&Clipping=showall

Lyotard, J. F. (1971). *Discourse, figure*. Paris: Klincksieck.

Mahizhnan, A., & M. T. Yap (2000). Singapore: The development of an intelligent island and social dividends of information technology. *Urban Studies, 37*(10), 1749–1756.

Miller, D., & D. Slater (2000). *Internet: An ethnographic approach*. Oxford: Berg.

Nee, S. C. (July 17, 2005). Fast and furious reaction to NKF revelations. *Star*. Retrieved September 17, 2007, from http://www.singapore-window.org/sw05/050717st.htm

Nardi, B. A., Diane J Schiano, Michelle Gumbrecht, Luke Swartz (2004). Why we blog. *Communications of the ACM, 47*(12), 41–46.

Parliament Sitting (April 3, 2006). *General election: Podcasting for campaigning not allowed* (Question No. 424 for Oral Answer). Available from Ministry of Information, Communication and the Arts Web site: www. mica.gov.sg

Rheingold, H. (1994). *The virtual community*. New York: HarperPerennial.

Riffattere, M. (1990). Compulsory reader response: The intertextual drive. In Worton, M. & Still, J. (Eds.), *Intertextuality, theories and practices*. Manchester, UK: Manchester University Press, 56–78.

Rodan, G. (1998). The internet and political control in Singapore. *Political Science Quarterly, 113*(1), 68–75.

Rodan, G. (2000). Singapore: Information lockdown, business as usual. In L. Williams & R. Rich (Eds.), *Losing control: Freedom of the press in Asia* (pp. 169–189). Canberra: Asia Pacific Press.

Rodan, G. (2003). Embracing electronic media but suppressing civil society: Authoritarian consolidation in Singapore. *Pacific Review, 16*(4), 503–524.

Silverstone, R. (1999). *Why study the media?* London: Sage.

Siew, A., & Chua, H. (2006, May 4). Citizen reporting' despite ban on podcasts: Net abuzz with video clips, chatter, even a satirical podcast. *The Straits Times*.

South China Morning Post (2005, August 18). *Singapore net controls Shun filter*. Retrieved June 6, 2006, from http://www.asiamedia.ucla.edu.print.asp?parentid=282518

Stevens, J. E. (2006). Taking the big gulp: The web is its own medium with its own characteristics. It is not newspapers. It's not TV news. It is not radio. *Nieman Reports, 60*(4), 66–70.

Su, N. M., Yang Wang, Gloria Mark, Tosin Aiyelokun, Tadashi Nakano et al. (2005, June). *A bosom buddy afar brings distant land near: Are bloggers a global community*. In Proceedings of the Second International Conference on Communities and Technologies, Milan.

Sussman, G. (2003). The Tao of Singapore's internet politics: Towards digital dictatorship or democracy? *The Journal of International Communication, 9*(1), 35–51.

Sydney Morning Herald (2005, March 21). *More curbs on Singapore*. Retrieved June 23, 2007, from http://www.smh.com.au/news/breaking/more-singapore-curbs-on-blogs/2005/11/21/1132421574687.html#

Taipei Times (2005, September 19). *Singapore, Malaysia warn bloggers*. Retrieved June 6, 2005, from http://www.asiamedia.ucla.edu/print.asp?parentid=29966

Tan, T. (2006, April 7). Chances poor that public will take to the net during the polls. *The Straits Times*.

Tsang, S. (2001, October 18). *Singapore unveils rules for e-campaigning*. Retrieved July 20, 2005, from http://asia.cnet.com/news/industry/0,39001143,38025935,00.htm

Vasil, R. (2000). *Governing Singapore: A history of national development and democracy*. Singapore: Allen Unwin and Institute of Southeast Asian Studies.

Worthington, R. (2003). *Governance in Singapore*. London: RoutledgeCurzon.

Wills, C. (1989). Upsetting the public: Carnival, hysteria and women's texts. In K. Hirschkop & D. Shepherd (Eds.), *Bakhtin and cultural theory* (pp. 130–151). Manchester and New York: Manchester University Press.

Contributors

Debra Adams is a Ph.D. candidate in the Creative Industries Faculty at the Queensland University of Technology, Brisbane, Australia. She is looking at the capacity of new media to re-engage citizens in public affairs and democratize journalism practice. Adams holds a first class honors degree in Journalism from the Queensland University of Technology. Her work has appeared in *Journalism, Media and Communication*, and *Journalism Education*.

Karina Alexanyan is a Ph.D. Candidate in Communications at Columbia University, New York. Her doctoral research centers on the "role of global communication technologies in cultural globalization," with a focus on Russian-language Internet users. Alexanyan received her M.Phil. from Columbia University, has an M.A. in Communication from NYU and a B.A. in Linguistics and Modern Languages (French and Russian) from the Claremont Colleges.

Axel Bruns is Senior Lecturer in the Creative Industries Faculty at Queensland University of Technology in Brisbane, Australia. He is the author of *Blogs, Wikipedia, Second Life and Beyond: From Production to Produsage* (2008) and *Gatewatching: Collaborative Online News Production* (2005), and the editor of *Uses of Blogs* with Joanne Jacobs (2006; all released by Peter Lang, New York). He blogs at http://snurb.info/. Bruns has coined the term "produsage" to better describe the current paradigm shift toward user-led forms of collaborative content creation that are proving to have an increasing impact on media, economy, law, social practices, and democracy itself. For more information about the produsage concept, see Produsage.org.

Kim De Vries is Director of Composition and Assistant Professor of English at California State University, Stanislaus. She earned her M.A. and Ph.D. at the University of Massachusetts Amherst in English (concentration in Rhetoric and Composition). Her research interests center on the rhetorical construction of identity and community, particularly in contexts involving ICTs or new media. Since 2000, she has been a staff writer for Sequential Tart, a webzine devoted to comic books and pop culture. Recent publications include "Desire, Dissent and Differentiation: Sustaining Growth in Virtual Networks," *New Network Theory Reader*, Amsterdam, 2007 and "Writing Wonder Women: How Playful Resistance Leads to Sustained Authorial Participation at Sequential Tart," Forthcoming in *Writing and the Digital Generation*, Heather Urbanski, ed., McFarland, 2009.

Aziz Douai is Assistant Professor of International Communications at Franklin College, Switzerland, and a doctoral candidate in Mass Communications at Pennsylvania State University. Douai holds an M.Sc. in Advertising and Communication Research from Boston University. His research interests include social and political implications of mass media, new media and free speech, and the political economy of the media. A specialist in international communications, his doctoral dissertation focuses on international broadcasting and the management of foreign public opinion, particularly with respect to the case of Al-Hurra and Arab television. His latest publication, "Tales of Transgression or Clashing Paradigms: The Danish Cartoon Controversy and Arab Media," appeared in *Global Media Journal*.

Nabil Echchaibi is Assistant Professor in the School of Journalism at the University of Colorado-Boulder. His research is concerned with race and identity politics in Western Europe and diasporic media produced by Muslim diasporas. His most recent work has appeared in various international journals including *Javnost/The Public, Gazette: Journal of International Communication* and *The Journal of Intercultural Studies*. He is the author of *Voicing Diaspora: Ethnic Radio in Paris and Berlin between Cultural Renewal and Retention* (2009) with Lexington Books. He holds a Ph.D. in Communication from Indiana University.

Yasmin Ibrahim is a Senior Lecturer in the division of Information and Media Studies at the University of Brighton where she lectures on globalization and political communication. Her main research interests include the use of the Internet for empowerment and political communication in repressed polities and diasporic communities, global governance, and the development of alternative media theories in non-Western contexts. Her work has appeared in the *International Journal of Technology, Knowledge and Society*, the online journal RE-public, and in the book *Linguistic and Cultural Online Communication Issues in the Global Age* edited by K. St. Amant.

Olessia Koltsova is Associate Professor of Mass Communications at the State University Higher School of Economics in St. Petersburg, Russia. She holds a Ph.D. in Sociology from St. Petersburg State University. Over the past ten years she has been researching post-Soviet media transformation and development. She is the author of *News, Media and Power in Russia*, (Routledge, 2006).

Giovanni Navarria is Research Associate at the Centre for the Study of Democracy (CSD), at the University of Westminster, London, where he is also completing a Ph.D. on the future of political activism in the age of social and communication media networks. Navarria holds a *Laurea in Filosofia* from the University of Catania. His latest publication is a chapter titled "A New Technology of Control: E-Government," in *E-Government: Opportunities and Challenges in Developed Nations*. ICFAI Press.

Adrienne Russell is Assistant Professor of Digital Media Studies at the University of Denver. Her research centers on emerging media tools and practices and how they impact activist media and journalism. Her work has appeared in peer-reviewed journals including *Critical Studies in Media Communication, New Media and Society*, and *Journalism: Theory, Practice, and Criticism*. She is also a contributor to the following books: *The Anthropology of News and Journalism: Global Perspectives* (Indiana University Press, 2009) and *Networked Publics* (MIT Press, 2008). She holds a Ph.D. in Communication from Indiana University.

Eugenia Siapera is Lecturer in Media and Communication at the University of Leicester, and Course Director of the M.A. New Media and Society. She holds a Ph.D. in Social and Political Sciences from the European University Institute in Florence, Italy. Siapera has written on aspects of new media and multiculturalism, and her articles have appeared in several journals, including *New Media and Society, Journal of Ethnic and Migration Studies*, and *The European Journal of Cultural Studies*. She has also co-edited two books: *At the Interface: Continuity and Transformation in Culture and Politics* (with Joss Hands, Rodopi, 2004) and *Radical Democracy and the Internet* (with Lincoln Dahlberg, Palgrave, 2007).

Carmel L. Vaisman is a doctoral student in the Department of Communications of the Hebrew University of Jerusalem, writing her doctoral dissertation on the Hebrew blogosphere. She received her graduate degree in Professional Communications from Clark University in Massachusetts and her undergraduate degree in political science, sociology, and anthropology from the Hebrew University of Jerusalem. Her work on the Hebrew-language Internet has been published in the popular and academic press in both Hebrew and English.

Index

General Editor: *Steve Jones*

Digital Formations is an essential source for critical, high-quality books on digital technologies and modern life. Volumes in the series break new ground by emphasizing multiple methodological and theoretical approaches to deeply probe the formation and reformation of lived experience as it is refracted through digital interaction. **Digital Formations** pushes forward our understanding of the intersections—and corresponding implications—between the digital technologies and everyday life. The series emphasizes critical studies in the context of emergent and existing digital technologies.

Other recent titles include:

Leslie Shade
  Gender and Community in the Social Construction of the Internet

John T. Waisanen
  Thinking Geometrically

Mia Consalvo & Susanna Paasonen
  Women and Everyday Uses of the Internet

Dennis Waskul
  Self-Games and Body-Play

David Myers
  The Nature of Computer Games

Robert Hassan
  The Chronoscopic Society

M. Johns, S. Chen, & G. Hall
  Online Social Research

C. Kaha Waite
  Mediation and the Communication Matrix

Jenny Sunden
  Material Virtualities

Helen Nissenbaum & Monroe Price
  Academy and the Internet

To order other books in this series please contact our Customer Service Department:
  (800) 770-LANG (within the US)
  (212) 647-7706 (outside the US)
  (212) 647-7707 FAX

To find out more about the series or browse a full list of titles, please visit our website:
  WWW.PETERLANG.COM